BIOCHEMISTRY RESEARCH TRENDS

PROTEASES

FUNCTIONS, MECHANISMS AND USES

BIOCHEMISTRY RESEARCH TRENDS

Additional books and e-books in this series can be found on Nova's website under the Series tab.

BIOCHEMISTRY RESEARCH TRENDS

PROTEASES

FUNCTIONS, MECHANISMS AND USES

NELSON PÉREZ GUERRA
EDITOR

nova
science publishers
New York

Library of Congress Cataloging-in-Publication Data

ISBN: 978-1-53615-854-0
Library of Congress Control Number:2019944889

Published by Nova Science Publishers, Inc. † New York

Enzymes are strategic tools for the green industry and, among them, proteases are of main interest because of their wide applications. This book about proteases brings together basic knowledge about sources, classification and mechanisms of reaction of these enzymes, and particular and new concepts regarding proteases of marine, protozoa or vegetal origin, proteases implications in agriculture, and powerful tools related to directed evolution, metagenomics and bioinformatics applied to the screening, development and study of new proteases with enhanced properties. For all that, this book is a very useful and comprehensive source of knowledge and information for students, academics and biotechnologists who are interested in the study, research and application of these enzymes.

Ana Torrado Agrasar (PhD)
Professor/Researcher
Faculty of Sciences, University of Vigo, Ourense (Spain).

CONTENTS

PREFACE

Proteases or proteinases are a group of enzymes that catalyze the hydrolytic breakdown of proteins produced by animals, plants and microorganisms. These enzymes have a wide application in several industrial processes including cheese production, meat tenderization, production of protein hydrolysates, silver recovery and silk; in many other fields including baking, textiles, pharmaceutical and leather industries and also, as an additive to detergents and in bioremediation procedures.

However, for an appropriate and economical industrial application of protease enzymes, it is necessary to know their functions, origin (plants, animals, protozoa, bacteria and fungus), classification and characteristics, mechanisms of action, their specific target substrates and products obtained in the reactions that these enzymes catalyze, as well as their optimal temperature and pH of activity and stability.

Therefore, the aim of this book is to provide an overview of the latest knowledge about the functions, mechanisms and applications of protease enzymes in different industrial fields. I hope that the content of this book will provide useful information to doctors, biochemists, pharmacists, biologists, microbiologists and food technologists on the practical knowledge accumulated by those researchers who have worked in the field of these enzymes. The information provided in this book can also serve as a concentrated bibliographic source for academic purposes for professors,

PhD students and Food Science and Technology, Biochemistry, Biology, Medicine, Pharmacy undergraduate students.

The book includes chapters related with: i) the role and different structures, mechanism, composition and specificity of proteases (chapter 1), ii) the different origin, characteristics and action of insect digestive proteases (chapter 2), marine proteases (chapter 3) and proteases from protozoa (chapter 4), iii) the emerging technologies developed to produce novel proteases with enhanced specificity, stability and catalytic activity for a wide industrial application (chapter 5), iv) the application of proteases to obtain functional foods, *i.e.*, sweet potato protein hydrolysates (chapter 6), v) the inhibition of protease activity by natural protease inhibitors including trypsin and chymotrypsin inhibitors (chapter 7), and vi) the use of bioinformatics resources (databases, tools, visualization and data analysis) to provide a better understanding of proteases and their functions at the molecular level (chapter 8).

Chapter 1 - Considered as principal mediators of catalysis of enzyme and composition, objectives for drug advance, or chemical agents for green chemistry, proteases have extensively commanded the captivated consideration of doctors, biochemists, and manufacturers. The ubiquity and effect of proteases in nature are often unmatched. A small number of genes in the majority organisms are proteases, second in number merely to transcription factors, and proteases are implied in virtually every physiological operation.

Proteases are known as the hydrolyzing enzymes that catalyze the hydrolysis of the peptide bonds in the primary configuration of peptides and proteins. They are commonly used to cut the proteins and then particularly to create new functional peptides in the different processes. Proteases are detected in a large assortment of living organisms and they as well demonstrate diverse chemical, physicochemical, biological and physiological roles in the world. They are considered as the most important enzymes in the engineering, accounting for more than half of the total enzyme scales in the globe. Proteases are the hydrolyzing enzymes which have limited size, condensed molecules, and uncomplicated structures. They are used in the regularization and moving of metabolism, gene expression, enzyme change and the processes of the production.

This review gives a general overview of the role and diversity of the structure of proteases, including mechanism, composition, and specificity.

Chapter 2 - Peptidases are hydrolases, enzymes that are able of cleaving peptide bonds between amino acids residues. These enzymes can be classified according to the position of the bond they hydrolyze (i.e., endo or exopeptidases), catalytic site (i.e., aspartic or cysteine proteinases), or their mode of action. They may also be classified based on their primary structure, into clans and families within major classes (i.e., Family A1 or C1). Proteins are one of the most important components of insect food intake. Therefore, peptidases are fundamental in the breakdown processes of food macromolecules into smaller molecules, allowing absorption of nutrients during digestion. Peptidase enzymology and the physiological understanding of such proteins are significant fields of study due to the financial and health bearings in which many insect species are involved, acting as pests for crops or vectors for causative agents of diseases. In this way, peptidases are essential targets for the development of inhibitors focusing on the prevention of herbivory, blood digestion in hematophagous insects or even in the invertebrate host-parasite interactions. Four major peptidase groups accomplish insect protein digestion. Serine proteinases, such as trypsins and chymotrypsins, are common to most herbivore and phytophagous species as well as the majority of the hematophagous insects; cysteine and aspartic proteinases, generally found in association, such as cathepsins D and L-like enzymes in hemipterans; and metallopeptidases, as is the case of several aminopeptidases. In this chapter, we will comment on the most recent aspects in the study of protein digestion in insects, detailing particularities of larval and adult main enzymes, adaptations for specific food digestion, and evolutionary tendencies.

Chapter 3 - Organisms inhabiting the marine environments are excellent sources of novel enzymes, which retain their integrity and function under extreme or different reaction conditions. Proteolytic enzymes, or proteases, are a group of enzymes able to hydrolyze protein peptide bonds. These are classified by several criteria. By their optimal pH, they are neutral, acidic or alkaline. With respect to their active center, enzymes are divided into serine, cysteine, aspartyl, and metallo-proteases. Marine proteases have unique properties, mainly high activity at low-temperatures, thermal stability and pH adaptation. Novel enzymes with particular characteristics or functions are valuable in industrial bioprocessing. In this respect, functions and mechanisms of marine proteases are similar to proteases from microorganisms, plants, and warm-blood animals, although their application can be in wider processing conditions, the action mechanism of specific marine proteases are studied at present by several in silico analyses. Marine protease applications are extensive. In aquaculture, in developing diets to several species. In the food industry, proteases are widely used for meat tenderization, fat emulsifying, foaming, beverage clarification, debittering, texturization, deproteinization and flavor production. In therapeutics, biopeptides produced by hydrolyzing specific proteins act as antioxidants, anticoagulants, antihypertensive, antitumor, antinflammation and anticancer, among others. Collagen obtained from marine sources using marine enzymes is also used in several industries, such as cosmetics, foods and medicine. Additionally, collagen hydrolysates are also bioactive (antioxidant, antimicrobial, antifatigue, anticancer, with mineral and hormonal regulating properties and angiotensin I converting enzyme or ACE). In addition, bioactive peptides and enzymes are also obtained from fish and seafood wastes such as shells, viscera, heads, skins, bones, fins and trimmings, helping to alleviate a serious pollution problem, as around 75% part of fish is discarded as a waste.

Chapter 4 - Parasitic protozoa are capable of causing both acute and chronic infections in humans; as a whole, they are estimated to affect over 500 000 people worldwide, causing malaria, trypanosomiasis, amoebiasis, leishmaniasis, giardiasis, and trichomoniasis, among other important diseases. To be able to occupy different niches, parasites have developed different strategies to evade the immune response of the host. Within these strategies is the use of secretion proteinases that are able to degrade different components of the innate immune system, such as mucus, which is found mainly in respiratory, digestive and vaginal epithelium, and other proteins, such as degradation of MUC5A; Acanthamoeba castellanii degrades lactoferrin, while Trichomonas vaginalis has surface and secreted proteinases that degrade mucus and allow it to adhere to epithelial cells. For researchers and clinicians, it is important to understand how protozoan proteases act against the different molecules of the innate immune system to develop strategies that allow the blockage of these molecules, such as the design of specific antiparasitic drugs that impede colonization and imminent parasitism.

Chapter 5 - With regards to massive industrialization, the demand for enzymes promotes rapidly. Proteases as unique classes of enzymes have various roles in industries. However, some of them are hindered by low catalytic activity and insufficient stability during the different industrial process. Therefore, researchers try to improve the ability of these biocatalysts through the discovery of novel enzyme with higher functionality or increased properties through new technology including protein engineering, metagenomics, immobilization, etc. This chapter is reviewed the emerging technologies for the development of novel industrial proteases.

Chapter 6 - This chapter introduces the overview, preparation and functional properties of hydrolysates from sweet potato protein (SPPH), as well as its antioxidant and antihypertensive activities. It starts by presenting the overview of SPPH, followed by the preparation of SPPH as affected by hydrolysis condition, single and combined enzymes used. It then explains functional properties of SPPH as affected by pH, salts and high hydrostatic pressure (HHP). Then, antioxidant activity of SPPH is introduced, as well as the identification and characterization of its peptides with antioxidant potential. By the end of the chapter, antihypertensive activity of SPPH is presented in vitro and in vivo level.

Chapter 7 - Some of the constituents of food or mixed-feed have not yet been properly utilized because of the presences of anti-nutritional factors (ANF). One of the ANF that are commonly and varyingly distributed throughout the nature is called protease inhibitors (trypsin inhibitor and chymotrypsin inhibitor). Trypsin inhibitor (TI) is almost negligible in cereals, but highly concentrated in legume grains, particularly in soybeans. The TI is mainly located in the protein bodies of the cotyledon. The ANF can generally be grouped as heat-labile, heat-resistant or heat-tolerant.

Depending on the processing conditions, certain amounts of TI is always available in the consumable products, for example, a soybean-protein isolate contain up to 40% of the TI activity. Bowman-birk inhibitor (BBI), a rich in the contents of cysteine has more resistance to heat, compared to Kunitz trypsin inhibitor (KTI). Methionine, cysteine, homo-cysteine, and taurine are the four common sulphur-containing amino acids (AA), but only the first two are naturally incorporated into proteins. A protein from soybean is not an ideal because of inadequacy of methionine (Met), but the BBI and to some extent KTI are positively contributing the quality due to their higher contents of cysteine. Digestibility of methionine could not be affected by the availability of TI, in the raw soybeans. The TI is generally responsible to negatively affecting endogenous trypsin enzymes and thereby creates pancreatic hypertrophy of some animals. However, there is uncertainty with regards to the biological threshold levels of this TI, in foods/diets of susceptible animals. Unlike cysteine the sulphur of methionine is not highly nucleophilic, and therefore, methionine from unheated soybean would slowly be liberated and absorbed. Although methionine is always an initiator to synthesis that of all types of eukaryotic proteins, cysteine has ability to form disulfide-bonds and also plays a crucial role of protein-structuring. It is therefore concluded that despite the fact that their availability certainly enriches the sulphur-containing amino acids, protease inhibitors are generally and usually impairing the nutritional values of most legume grains.

Chapter 8 - Bioinformatics is an interdisciplinary area that is associated with the development of various tools for analysis and interpretation of biological data. One of the major objectives of this chapter is to provide a better understanding of protease and its functions at the molecular level by using bioinformatics approaches including databases, tools, visualization and data analysis. Proteases are responsible for proteolysis of proteins, one of the major group of industrial enzymes involved in digestion, conception, birth, growth, maturation, aging and even death of all organisms. In addition to food industry, use of proteases in pharmaceutics is gaining more attention as several clinical studies are indicating their applications in various diseases. Therefore, the major focus of the present chapter will include many applications of tools and databases that are available for providing information on proteases like MEROPS, Gypsy Database, Knottin database, HIV protease database and PLANT-PI. The chapter gives a detailed overview of the significant database which provides information on enzymes such as BRENDA, KEGG and CSA. The present chapter also highlights ExploreEnz, IntEnz, EzCat database, SFLD and CPDB. The last section describes the tools like PROSPER, Peptide Cutter, PSORT, for enzyme analysis which will help the readers to effectively use the information for their research endeavors.

In: Proteases: Functions, Mechanisms and Uses ISBN: 978-1-53615-854-0
Editor: Nelson Pérez Guerra

Chapter 1

PROTEASES: USEFUL ENZYMES FOR LIFE

Karim Ennouri**, PhD
University of Sfax, Sfax, Tunisia

ABSTRACT

Considered as principal mediators of catalysis of enzyme and composition, objectives for drug advance, or chemical agents for green chemistry, proteases have extensively commanded the captivated consideration of doctors, biochemists, and manufacturers. The ubiquity and effect of proteases in nature are often unmatched. A small number of genes in the majority organisms are proteases, second in number merely to transcription factors, and proteases are implied in virtually every physiological operation.

Proteases are known as the hydrolyzing enzymes that catalyze the hydrolysis of the peptide bonds in the primary configuration of peptides and proteins. They are commonly used to cut the proteins and then particularly to create new functional peptides in the different processes. Proteases are detected in a large assortment of living organisms and they as well demonstrate diverse chemical, physicochemical, biological and physiological roles in the world. They are considered as the most important enzymes in the engineering, accounting for more than half of the total

* Corresponding Author's E-mail: 1karimennouri1@gmail.com.

enzyme scales in the globe. Proteases are the hydrolyzing enzymes which have limited size, condensed molecules, and uncomplicated structures. They are used in the regularization and moving of metabolism, gene expression, enzyme change and the processes of the production.

This review gives a general overview of the role and diversity of the structure of proteases, including mechanism, composition, and specificity.

Keywords: proteases, enzyme, peptide, protein, catalysis

1. INTRODUCTION

Proteases are one of the biggest and most various families of enzymes recognized to date. These are as well called as proteinases and peptidases. Proteases are mainly judged as the enzymes of digestion. These transform proteins into small peptides or amino acids. These are principally categorized under Group 3 (Hydrolases) and sub-group 4 on the basis of the classification commission of the International Union of Biochemistry and Molecular Biology. Proteases do not limit to only one enzyme although a combination of enzymes comprising amidases and proteinases [1]. These enzymes have multipurpose functions in different manufacturers, such as production of protein hydrolysates, silver recovery, silk, food, leather, pharmaceutical, and detergents [2]. Proteases are in addition conceived as having wide appliances in progress of "green" and eco-friendly expertise as well as in a number of bioremediation procedures [3, 4]. The world market for industrial enzymes has an average growth rate of approximately 10% per year [5]. Microbial proteases represent one of the most significant factions of manufacturing enzymes, supplying for more than 65% of total manufacturing enzyme market [5]. The domination of proteases in the manufacturing marketplace is expected to enhance more in the next years. Among different types (acidic, neutral and alkaline) of proteases, alkaline proteases are the most commonly used industrial enzyme in view of their activity and stability at high pH [6, 7].

2. Proteases: Origins and Characteristics

Proteases are acquired from large selections of sources such as microorganisms, animals, and plants. A microbial assortment is a preferred basis of proteases for the reason that they have a broad biochemical diversity, the achievability of mass culture and effortlessness of genetic management. The bacteria can generate a plentiful and usual stock of the desired molecules and can be produced in large amounts in a reasonably short amount of time by showing fermentation techniques. Additionally, proteases from bacteria have improved shelf-life and can be stocked under less than adequate states for several weeks with no considerable failure of action. A wide diversity of microorganisms like actinomycetes, fungi, bacteria and also yeast are well-known to generate these enzymatic proteins [8]. Besides, proteases from numerous bacteria have been characterized and purified from diverse sources. The majority of industrial alkaline proteases are generated by microorganisms, in particular *Bacillus* sp. for the reason of their elevated pH and thermal constancy [9]. A number of *Bacillus* species well recognized for protease making are *Bacillus subtilis*, *Bacillus megaterium, Bacillus amyloliquifaciens*, *Bacillus mojavensis*, *Bacillus cereus*, *Bacillus thuringiensis* and also *Bacillus sterothermophilus* [10, 11]. Microbial proteases are of two different categories, extracellular and intracellular [3]. Intracellular proteases are significant for an assortment of metabolic and cellular procedures like for example preservation of the protein pool in the cell, maturation of hormones and enzymes, protein turnover, sporulation, and segregation, whereas extracellular proteases are essential for the protein hydrolysis in cell-free areas which facilitate the cell to attract and exploit hydrolytic products. Besides, extracellular proteases are frankly exuded into the fermentation medium by the manufacturer, consequently making simpler the downstream processing of the enzyme as contrasted to proteases acquired from animals or plants [12]. In addition, enzymes are needed for digesting nutrients and enable the cell to absorb hydrolyzed products in order to conclude microbial growth stages [13].

Protease production through microorganisms is deeply controlled by media constituents, particularly nitrogen and carbon supplies and physical features like for example temperature, inoculum density, pH, agitation and incubation time [14, 15, 16]. Moreover, Proteolytic enzyme production by *Bacillus* is highly influenced by media components as carbon, nitrogen, the presence of easily metabolizable sugars [17] and metal ions [18]. Li and Yousten [19] concluded that the proteolytic activity produced by *B. thuringiensis subsp. kurstaki* was dependent upon supplementation with Mn^{2+}. It was reported that the addition of Mg^{2+} and K^{+} salts to the culture medium resulted in high production of protease [20, 21]. It is also known that carbon, nitrogen and sulfate sources contribute to alkaline protease production from *Bacillus* sp. [22]. In addition, recent statistical experimental designs such as response surface methodology and investigation factorial design have been employed for bioprocess optimization taking into consideration the interactive influences in the midst of the control feature for overproduction of proteases [23, 24, 25]. Bacterial proteases are generally produced by both immobilized and free cells. Immobilized entire cell/enzyme assembly suggests a number of benefits such as greater effective constancy due to biocatalyst and elevated effectiveness of catalysis. The control of an enzyme permits for an uncomplicated breakup of the enzyme from the products and for reusing of the enzymatic protein [26]. Immobilized cells are employed in agro-alimentary technology, analytical chemistry, biomedicine and biotechnology [27]. It attempts a range of benefits, such as rise of production due to the elevated cell amount within the bioreactor, anticipation of failure in a continuous manoeuvre in the midst of others [28], and capability to detach cell mass from the bulk fluid for probable reuse [27, 29].

Nevertheless, proper selection of immobilization methods and supporting types of equipment is wanted to diminish the drawbacks of control. Normally, enzymatic proteins or cellular organisms are stopped by cross-linking physical adsorption, ionic or covalent binding, and entrapment techniques. Entrapment technique has been judged most appropriate among a variety of reported techniques. It comprises of enclosing the enzymatic protein or cell in an aqueous solution within a semi-porous membrane container. In this restrictive practice, the particle configuration permits contact between the substrate and enzymatic protein (or cell) to be completed, and it is feasible to put out of action in a number of enzymatic proteins (or cells) at the same moment [26].

The greatest appliance of alkaline proteases is in washing cleansers, where they facilitate in eliminating protein based marks from clothing and on the cover dehairing method, where dehairing is done at pH ranged between 8 and 10. For an enzyme to be employed as a cleanser stabilizer, it should be steady and active in the presence of classic detergent components as an example of surfactants, constructors, cloth softeners and other several formulation supports [30].

Proteases are also used in the silver recuperation appliance. The dregs of X-ray photographic films, including black metallic silver extend in gelatin are a very important supply of silver and polyethylene terephthalate upturn [31]. The quantity of silver in this type of layers changes between 1.5 to 2.0 percent by mass [32]. The silver revival techniques include straight flaming of a layer, oxidation of metallic silver trailed by electrolysis, banding the silver substance coat via microbial enzymatic proteins, in particular alkaline proteases and the acid trickling procedure [33, 34]. On the other hand, more to the point of creating ecological contamination and health risks, the burning of films provides pieces of evidence inefficient in polyethylene terephthalate recuperation [35]. For that reason, there is an imperative call to build up lucrative and toxic waste free techniques. The hydrolysis of the gelatin layers on the X-ray coat by proteolytic enzymes of microbial origins does not simply make possible the revitalization of silver although furthermore polyethylene terephthalate base, which can be reused. Nevertheless, just a small number of studies with microbial proteases have been described as a substitute for the burning and oxidation techniques of silver recuperation [36].

There is an extraordinarily elongated list defining the diversity of bacteria engaged in protease secretion. However, despite such an extensive list, just a small number is judged as suitable protease makers for industrial use, is normally regarded as secure, non-dangerous and non-pathogenic. As a consequence, the exploration of novel microbial origins of proteases is a currently happening practice. The bacteria from varied and exotic locations named extremophiles are judged as a significant origin of enzymatic proteins and their particular characteristics are supposed to result in new process appliances. At present, rDNA knowledge is increasing in use to expand genetically transformed microorganisms for enhanced fabrication of protease. In fact, separation, duplicating and expression of genetic material of protease symbolize an indispensable action in the manufacturing of the most part of competent microbes [37]. More than a few studies and reports have demonstrated the fabrication, refinement, limited description, molecular duplicating, sequencing of alkaline protease [38, 39, 40]. Similar investigations are indispensable to fabricate more proficient protease manufacturer, which permits enhancement of its different appliances in diverse industries.

On another hand, rhizosphere is known as the part of the soil that surrounds the roots of plants and is transformed by the plant's root growth, nutritive components, respiration, etc. Rhizosphere symbolizes an immense pool of possible microbial variety and presents an under-investigated origin to monitor adaptable microorganisms with natural steady proteases. Rhizosphere bacteria, generally, have an essential function that positively influences plant development and productivity [41]. Plant development promoting rhizobacteria (PDPR) have been deeply learned for the reason that they have the aptitude to support plant development, improve crop efficiency and as possible biocontrol instruments. Beside numerous other types of action of PDPRs responsible for the inhibition of plant pathogens, newly consideration has been focused to the fabrication of lytic enzymatic proteins due to the fact that they have important functions in breaking down the cellular barriers of pathogens and restraining the spore germinations [38, 42]. The cellular fence-degrading enzymes, mainly glucanases, chitinases, and also proteases, are most important lytic enzymatic proteins that are secreted by means of biocontrol mediators [43]. Production of proteases by PDPRs having opposed action are in recent times intensively investigated for sustainable harvest [44, 45], on the other hand, the function of proteolytic PDPRs in a range of industries such as leather, cleanser and silver revival is even an unheard investigation [46].

Nowadays, there has been enhanced attention in proteases as objectives for expanding therapeutic mediators [47]. Proteases are omnipresent in the environment and found in an extensive assortment of microorganisms [48]. Protease fabrication is a natural property of all life forms, however, just those that secrete high-supply extracellular protease has been used commercially [49]. Proteases are molecules of reasonably small dimension and are condensed, globular structures that catalyze the peptide connection division in proteins. These enzymatic proteins are vital in a number of diverse and decisive biological processes; for instance, they are engaged in the hydrolysis of big proteins to slighter molecules for a biological move and metabolic processes [50]. Along with the proteases, microbial proteases are more important than animal or fungal proteases. *Bacillus* is the most known species producing extracellular proteases in the midst of a lot of microbial species that have found great appliance in pharmaceutical, food processing engineering, laundry and leather [3, 51].

Proteases carry out a huge diversity of purposes and cover main biotechnological claims. They correspond to one of the major groups of manufacturing enzymes and come across an appliance in pharmaceutical production, detergent production, food production, leather production, and bioremediation procedures. Almost certainly the largest use of proteases is in laundry detergents, where they facilitate eliminating protein based marks from wrapping. In cloth production, proteases might besides be employed to eliminate the rigid and dry gum coat from the untreated silk fiber to attain developed brilliance and elasticity. They are capable of changing the exterior of silk and wool fibers to offer a novel and exclusive end. Proteases have been employed in the pelt-dehairing method, in which dehairing is effective at pH values betwixt 8 and 10 [52].

Alkaline proteases isolated in unpolluted shape from a number of microbial species have been revised widely to figure out their manufacturing significance. A specified revision of their characteristics similar to most favorable pH [53], heat [54], molecular dimension, a consequence of stabilizing mediators and a variety of restraining aspects provides a perception of the precise manufacturing appliance. On the other hand, it is a fully recognized statement that just steady biocatalysts able of enduring unkind circumstances of the procedure have a probability of business. Trade appliance of alkaline proteases claims for a higher point of acceptance in the direction of a range of physical circumstances (pH and temperature) and also chemical mediators (NaCl, detergents, metal ions, etc.) [55].

Proteases are difficult to characterize because of their diversity of action and structure. Originally proteases were classified based on molecular size, charge or substrate specificity. However, with the advent of molecular biology, proteases are now grouped into families based on the following; chemical nature of the catalytic or active sites, the mechanism(s) of action, and the evolutionary relationship of their three-dimensional structure [5].

3. Protease Classification

Terminology and classification are fundamentally imperative for information management. They permit individuals to converse well, with a total understanding of what they are chatting about and to collect and get back information well and clearly. A high-quality structure must have immense criteria to hide the whole flow of current data and be capable to rapidly take on the coming applicant in an acceptable position. There are three helpful ways of classifying proteases:

1) By the chemical process of catalysis.
2) By the site of action, mode and system.
3) By molecular organization and similarity.

3.1. Clustered by the Chemical Mechanism of Catalysis

Hartley [56] commenced the classification of peptide molecules on the foundations of their most important catalytic remains nature, that's why an extremely helpful theory came into view that supplementary turned out to be the support of the recent peptide categorization structure. This structure brought molecular compositions under an umbrella on the supports of catalytic systems, similar to threonine, cysteine, serine, glutamic, aspartic, or metallocatalytic peptidase types [56]. However, on the whole frequently they do not have any important homology to every other and have to be classified further.

3.2. Clustered by the Position of Action, Way and Mechanism

This plan has further classified proteases on the supports of arrangement related selectivity for the hydrolysis in an analogous cluster. On this foundation, they are capable to be organized into two main and two minor sub-classes, specifically endopeptidases (operating away from the end of the line of huge proteins), exopeptidases (operating on the end of the line of the polypeptide), oligopeptidases (operating away on the end of the line in smaller proteins) and in addition Omega-peptidases (operating on the end of the line of proteins) correspondingly. The two most important classes are in that case separated into sub-subclasses depending on their position and mode of action and somewhere in the foundation of the methodology of achievement [57]. This categorization was principally proposed by Enzyme Commissions' Nomenclature Committee [58, 59]. For this reason, this taxonomy has importance in common sense.

3.2.1. Endopeptidases

These enzymes take action on the alpha-peptide connections positioned away from the *N*-termination or *C*- termination. Some general illustrations are papain, chymotrypsin and pepsin. The endopeptidases comprise a precise and partial function in protein hydrolysis or proteolysis, similar to the elimination of the signal of peptides from exuded proteins and the transformation of the predecessor of proteins as enteropeptidase [60]. The endopeptidases are separated into sub-subclasses on the foundation of the catalytic system, and particularity is employed just to recognize special enzymes inside the sets. These are the sub-subclasses of threonine endopeptidases, serine endopeptidases, metalloendopeptidases, aspartic endopeptidases, and cysteine endopeptidases. Endopeptidases that could not be allocated in the least in the sub-subclasses were recorded in sub-subclass unassigned endopeptidases [61].

3.2.2. *Oligopeptidases*

Oligopeptidases tend to act on substrates smaller than proteins. Example of oligopeptidase is Thiamin-methionin oligopeptidase [62].

3.2.3. *Exopeptidases*

The exopeptidases have need of a liberated *N*-termination amino set, *C*-terminal carboxyl group or mutually, and hydrolyze a union of some rests from the termination (not more than three) [63]. The exopeptidases are more separated to *N*-termination operating peptidases, *C*-termination operating peptidases and dipeptidases. Every one of them is characterized as explain below.

3.2.3.1. *N*-Termination Operating Peptidases

The aminopeptidases refer to parts on the modified *N*-termination of its substrate and liberate a solitary amino acid rest. Models are aminopeptidase-*N* as well as aminopeptidase-*C*. Dipeptidyl-peptidase decomposes an *N*-termination dipeptide from the initial substrate. Models are dipeptidyl-peptidase I and dipeptidyl-peptidase III [64]. Tripeptidyl-peptidase, decomposes a tripeptide from the N-termination of the substrate. Models are tripeptidyl-peptidase I and also tripeptidyl-peptidase II [65].

3.2.3.2. C-Termination Operating Peptidases

These peptidases decompose a unique rest, from the free C-termination of the initial substrate. models are carboxypeptidase A1 and carboxypeptidase Y. Carboxypeptidases are separated in three subsubclasses: i) serine-type carboxypeptidases, ii) metallocarboxy-peptidases and iii) cysteine-type carboxypeptidases [66]. A peptidyl-dipeptidase decomposes a dipeptide from the *C*-termination of the cibled substrate: peptide+dipeptide. A model is peptidyldipeptidase A and angiotensin translating enzymatic protein [67].

3.2.3.3. Dipeptidases

These enzymatic proteins decompose a dipeptide, and usually need that both terminations are liberated. Models are dipeptidase A and dipeptidase of the membrane.

3.2.4. Omega-Peptidases

It is the sort of peptidases that haven't condition for an open *N*- or *C*-termination in the substrate. Nevertheless, their position of the action stays reserved for only a single terminal or another termination. As a result, they are completely different from both the endopeptidases and exopeptidases. They could be able to be active on the last residues that are replaced, sequenced or else related by means of isopeptide links. Isopeptide links are peptide relations erstwhile than those of a carboxyl to amino groups.

3.3. Clustered by Molecular Structure and Homology

This organization was previously expressed by Barrett and Rawlings [60]. It represents the most recent and managed organization that shows a relationship of individual unit commencing the evolutionary order and classifies them in similar protein sequence groups. Moreover, Rawlings and Barrett [68] explained a complex organization in which unique peptidases were allocated to sub-families of peptidases and similar peptidases starting diverse peptidic varieties combine to complete families, as well as families were more clustered in groups according to their three-dimensional composition analogy [68]. This idea was expanded to offer the constitution of the databases and enlarged to comprise the proteins that restrain peptidases [69, 70, 71]. This categorization scheme was original due to the fact that it has considered the entire of the protein into account that includes the peptide areas as their efficient entity; it rejects analogous areas with non-proteolytic efficient entities [72].

Conclusion

Microbial proteases constitute a complex and very large group of enzymes with great commercial value in food, detergent, leather, pharmaceuticals, and other industries. Different types of microbial proteases: alkaline, neutral and acidic have been characterized, purified and utilized till today. Looking into the commercial success of this enzyme class, researchers have now started aiming at the discovery and engineering of novel enzymes that are more robust with respect to their pH and temperature kinetics, using techniques of protein engineering and the identification of active site residues through chemical modifications, X-ray crystallographic data, and SDM. Researchers have started looking for newer hyper-producing strains and also exploited rDNA technology to develop genetically modified potential microorganisms for higher protease yield. Efforts are still being done to find newer enzyme sources and better production techniques. As their potential is much greater and their commercial value in the future is going to increase.

Hence, although microbial alkaline proteases already play an important role in several industries, their potential is much greater and their applications in future processes are likely to increase in the near future. The pursuit of other, newer approaches, such as novel discovery strategies targeting new dimensions of molecular diversity and technologies to improve performance characteristics by *in vitro* evolutionary changes of protein primary structures will certainly be the major field of development in next few years.

References

[1] Qureshi, A. S., Bhutto, M. A., Khushk, I. and Dahot, M. U. (2011). Optimization of cultural conditions for protease production by *Bacillus subtilis* EFRL 01. *African Journal of Biotechnology*, 10(26): 5173–5181.

[2] Annamalai, N., Rajeswari, M. V. and Balasubramanian, T. (2013). Extraction, purification and application of thermostable and halostable alkaline protease from *Bacillus alveayuensis* CAS 5 using marine wastes. *Food and Bioproducts Processing*, 92(4): 335–342.

[3] Jisha, V. N., Smitha, R. B., Pradeep, S., Sreedevi, S., Unni, K. N., Sajith, S., Priji, P., Josh, M. S. and Benjamin, S. (2013). Versatility of microbial proteases. *Advances in Enzyme Research*, 1(3): 39–51.

[4] Wang, S. L., Chao, C. H., Liang, T. W. and Chen, C.C. (2009). Purification and characterization of protease and chitinase from *Bacillus cereus* TKU006 and conversion of marine wastes by these enzymes. *Marine Biotechnology*, 11: 334–344.

[5] Bhunia, B., Basak, B. and Dey, A. (2012). A review on production of serine alkaline protease by *Bacillus* sp. *Journal of Biochemical Technology*, 3(4): 448–457.

[6] Denizci, A. A., Kazan, D., Abeln, E. C. A. and Erarslan, A. (2004). Newly isolated *Bacillus clausii* GMBAE 42: an alkaline protease producer capable to grow under highly alkaline conditions. *Journal of Applied Microbiology*, 96(2): 320–327.

[7] Devi, M. K., Banu, A. R., Gnanaprabhal, G. R., Pradeep, B.V. and Palaniswamy, M. (2008). Purification, characterization of alkaline protease from native isolate *Aspergillus niger* and its compatibility with commercial detergents. *Indian Journal of Science and Technology*, 1(7): 1–6.

[8] Sivasubramanian, S., Manohar, M. B. and Puvanakrishna, R. (2008). Mechanism of enzymatic dehairing of skins using a bacterial alkaline protease. *Chemosphere*, 70: 1025–1034.

[9] Rao, C. S., Sathish, T., Ravichandra, P. and Prakasham, R. S. (2009). Characterization of thermo and detergent stable serine protease from isolated *Bacillus circulans* and evaluation of eco-friendly applications. *Process Biochemistry*, 44: 262–268.

[10] Uyar, F., Porsuk, I., Kizil, G. and Yilmaz, E. I. (2011). Optimal conditions for production of extracellular protease from newly isolated *Bacillus cereus* strain CA15. *EurAsia Journal of BioScience*, 5: 1–9.

[11] Rani, M. R., Kumara, B. L., Sri, M. H. and Prasad, D. S. (2012). Isolation and screening of alkaline protease producing bacteria and induction of overproducing *Bacillus licheniformis* mutants through UV irradiation. *IOSR Journal of Pharmacy*, 1(1): 1–14.

[12] Haki, G. D. and Rakshit, S. K. (2003). Developments in industrially important thermostable enzymes: a review. *Bioresource Technology*, 89(1): 17–34.

[13] Padmaja, T., Suneetha, N., Sashidhar, R. B., Sharma, H. C., Deshpande, V. and Venkateswerlu, G. (2008). Degradation of the insecticidal toxin produced by *Bacillus thuringiensis var. kurstaki* by extracellular proteases produced by *Chrysosporium* sp. *Journal of Applied Microbiology*, 104: 1171–1181.

[14] Gupta, R., Beg, Q. K. and Lorenz, P. (2002). Bacterial alkaline proteases: molecular approaches and industrial applications. *Applied Microbiology and Biotechnology*, 59: 15–32.

[15] Abidi, F., Limam, F. and Marzouki, M. N. (2008). Production of alkaline proteases by *Botrytis cinerea* using economic raw materials: Assay as biodetergent. *Process Biochemistry*, 43(11): 1202–1208.

[16] Ennouri K., Ben Hassen H. and Zouari, N. (2013). Optimization of bioinsecticides overproduction by *Bacillus thuringiensis* subsp. *kurstaki* using linear regression. *Polish Journal of Microbiology*, 62 (3): 287–293.

[17] Gupta, R., Beg, Q. K., Khan, S. and Chauhan, B. (2002). An overview on fermentation, downstream processing and proprieties of microbial alkaline proteases. *Applied Microbiology and Biotechnology*, 60: 381–395.

[18] Varela, H., Ferrari, M. D., Belobrajdic, L., Weyrauch, R. and Loperena, M. L. (1996). Effect of medium composition on the production by a new *Bacillus subtilis* isolate of protease with promising unhairing activity. *World Journal of Microbiology and Biotechnology*, 12: 643–645.

[19] Li, E. and Yousten, A. A. (1975). Metalloprotease from *Bacillus thuringiensis*. *Applied Microbiology*, 30: 354–361.

[20] Ellaiah, P., Srinivasulu, B. and Adinarayana, K. (2002). A review on microbial alkaline proteases. *Journal of Scientific and Industrial Research*, 61: 690–704.

[21] Nadeem, M., Qazi, J. I., Baig, S. and Syed, Q. (2007). Studies on commercially important alkaline protease from *Bacillus licheniformis* N-2 isolated from decaying organic soil. *Turkish Journal of Biochemistry*, 32: 171–177.

[22] Chauhan, B. and Gupta, R. (2004). Application of statistical experimental design for optimization of alkaline protease production from *Bacillus* sp. RGR-14. *Process Biochemistry*, 39: 2115–2122.

[23] Saxena, R. and Singh, R. (2010). Statistical optimization of conditions for protease production from *Bacillus* sp. *Acta Biologica Szegediensis*, 54(2): 135–141.

[24] Ennouri, K., Ben Hassen, H., Ben Khedher, S. and Zouari, N. (2013). Concomitant production of proteases and delta-endotoxins of *Bacillus thuringiensis* subp *kurstaki* in a low-cost medium: effect of medium components. *Acta Biologica Szegediensis*, 57 (1): 13–19.

[25] Ennouri, K., Ben Ayed, R., Ben Hassen, H., Azzouz, H. and Triki M. A. (2015). Improvement of the Production of Entomopathogenic Proteases of *Bacillus thuringiensis*. *Tunisian Journal of Plant Protection*, 10(2): 95–104.

[26] Demirkan, E., Dincbas, S., Sevinc, N. and Ertan, F. (2011). Immobilization of *B. amyloliquefaciens* alpha-amylase and comparison of some of its enzymatic properties with the free form. *Romanian Biotechnological Letters*, 16(6): 6690–6701.

[27] Kumar, R. and Vats, R. (2010). Protease production by *Bacillus subtilis* immobilized on different matrices. *New York Science Journal*, 3(7): 20–24.

[28] Sankaralingam, S., Shankar, T., Sendeshkannan, K., Ramasubburayan, R. and Prakash, S. (2012). Production of Protease from *Pseudomonas* sp. by immobilization approach on different matrices. *European Journal of Applied Sciences*, 4(4): 146–156.

[29] Seabra, I. J. and Gil, M. H. (2007). Cotton gauze bandage: a support for protease immobilization for use in biomedical applications. *Brazilian Journal of Pharmaceutical Sciences*, 43(4): 535–542.

[30] Asker, M. M. S., Mahmoud, M. G., Shebwy, K. E. and Abd el Aziz, M. S. (2013). Purification and characterization of two thermostable protease fractions from *Bacillus megaterium. Journal of Genetic Engineering and Biotechnology*, 11: 103–109.

[31] Nakiboglu, N., Oscali, D. and Asa, I. (2000). Silver recovery from waste photographic films by an enzymatic method. *Turkish Journal of Chemistry*, 25: 349–353.

[32] Khunprasert, P., Grisdanurak, N., Thaveesri, J., Danutra, V. and Puttitavorn, W. (2008). Radiographic film waste management in Thailand and cleaner technology for silver leaching. *Journal of Cleaner Production*, 16: 28–36.

[33] Ajiwe, V. I. E. and Anyadiegwu, I. E. (2000). Recovery of silver from industrial wastes, cassava solution effects. *Separation and Purification Technology*, 18: 89–92.

[34] Cavello, I. A., Hours, R. A. and Cavalitto, S. F. (2013). Enzymatic hydrolysis of gelatin layers of x-ray films and release of silver particles using keratinolytic serine proteases from *Purpureocillium lilacinum* LPS#876. *Journal of Microbiology and Biotechnology*, 23(8): 1133–1139.

[35] Shankar, S., More, S. V. and Seeta Laxman, R. (2010). Recovery of silver from waste x-ray film by alkaline protease from *Conidiobolus coronatus. Kathmandu Unviversity Journal of Science and Engineering Technology*, 6: 60–69.

[36] Mausi, A., Yasuda, M., Fujiwara, N. and Ishikawa, H. (2004). Enzymatic hydrolysis of gelatin layers on used lith film using thermostable alkaline protease for recovery of silver and PET. *Biotechnology Progress*, 20: 1267–1269.

[37] Deesukon, W., Nishimurab, Y., Haradab, N., Sakamobob, T. and Sukhumsiricharta, W. (2011). Purification, characterization and gene cloning of two forms of a thermostable endo-xylanase from *Streptomyces* sp. SWU10. *Process Biochemistry*, 46: 2255–2262.

[38] Zhang, D., Spadaro, D., Valente, S., Garibaldi, A. and Gullino, M. L. (2012). Cloning, characterization, expression and antifungal activity of an alkaline serine protease of *Aureobasidium pullulans* PL5 involved in the biological control of postharvest pathogens. *International Journal of Food Microbiology*, 153: 453–464.

[39] Fang, D. A., Huang, X. M., Zhang, Z. Q., Xu, D. P., Zhou, Y. F., Liu, K., Duan, J. R., Shi, W. G. and Zhang, M. Y. (2013). Molecular cloning and expression analysis of chymotrypsinlike serine protease from the redclaw crayfish (*Cherax quadricarinatus*): a possible role in the junior and adult innate immune systems. *Fish and Shellfish Immunology*, 34: 1546–1552.

[40] Ennouri, K., Ben Hassen, H. and Zouari N. (2016). Statistical analysis of cultural parameters influencing delta-endotoxin and protease productions by *Bacillus thuringiensis kurstaki*. *Arabian Journal for Science and Engineering*, 41(1): 1–8.

[41] Garbeva, P., Van Veen, J. A. and Van Elsas, J. D. (2004). Microbial diversity in soil: selection of microbial populations by plant and soil type and implications for disease suppressiveness. *Annual Review of Phytopathology*, 42: 243–270.

[42] Ran, L. X., Liu, C. Y., Wu, G. J. and Van Loon, L. C. (2005). Suppression in bacterial wilt in *Eucalyptus urophylla* by fluorescent *Pseudomonas* sp. in China. *Biological Control*, 32: 111–120.

[43] Tseng, S., Liu, S., Yang, H., Lo, C. and Peng, K. (2008). Proteomic study of biocontrol mechanisms of *Trichoderma harzianum* ETS 323 in response to *Rhizoctonia solani*. *Journal of Agricultural and Food Chemistry*, 56: 6914–6922.

[44] Beneduzi, A., Ambrosini, A. and Passaglia, L. M. P. (2012). Plant growth promoting rizobacteria: their potential as antagonists and biocontrol agents. *Genetics and Molecular Biology*, 35(4): 1044–1051.

[45] Illakkiam, D., Anuj, N. L., Ponraj, P., Shankar, M., Rajendhran, J. and Gunasekaran, P. (2013). Proteolytic enzyme mediated antagonistic potential of *Pseudomonas aeroginosa* against *Macrophomina*

phaseolina. Indian Journal of Experimental Biology, 51(11): 1024–1031.

[46] Arunachalam, C. and Saritha, K. (2009). Protease enzyme: an eco-friendly alternative for leather industry. *Indian Journal of Science and Technology*, 2(12): 29–32.

[47] Shivanand, P. and Jayaraman, G. (2009). Production of extracellular protease from halotolerant bacterium, *Bacillus aquimaris* strain VITP4 isolated from Kumta coast. *Process Biochemistry*, 44: 1088–1094.

[48] Ennouri, K., Ayed, R. B., Triki, M. A., Ottaviani, E., Mazzarello, M., Hertelli, F. and Zouari, N. (2017). Multiple linear regression and artificial neural networks for delta-endotoxin and protease yields modelling of *Bacillus thuringiensis. 3 Biotech*, 7(3): 187.

[49] Cheng, S. W., Wang Y. F. and Wang, M. L. (2012). Statistical optimization of medium compositions for alkaline protease production by newly isolated *Bacillus amyloliquefaciens. Chemical and Biochemical Engineering Quarterly*, 26 (3): 225–231.

[50] Kumar, S., Sharma, N. S., Saharan, M. R. and Singh, R. (2005). Extracellular acid protease from *Rhizopus oryzae*: purification and characterization. *Process Biochemistry*, 40: 1701–1705.

[51] Ennouri, K., Ben Khedher, S., Jaoua, S., Zouari N. (2013). Correlation between delta endotoxin and proteolytic activities produced by *Bacillus thuringiensis* var. *kurstaki* growing in an economic production medium. *Biocontrol Science and Technology*, 23(7): 756–767.

[52] Ray, A. (2012). Protease enzyme- potential industrial scope. *International Jounal of Technology*, 2(1): 1–4.

[53] Martinez, R. H., Sanchez, G. G., Bergmann, C. W., Loera-Corral, O., Rojo-Dominguez, A., Huerta-Ochoa, S., Regalado-Gonzalez, C. and Prado-Barragán, L. A. (2011). Purification and characterization of a thermodynamic stable serine protease from *Aspergillus fumigatus. Process Biochemistry*, 46: 2001–2006.

[54] Jaouadi, B., Ellouz-Chaabouni, S., Rhimi, M. and Bejar, S. (2008). Biochemical and molecular characterization of a detergent-stable

serine alkaline protease from *Bacillus pumilus* CBS with high catalytic efficiency. *Biochimie*, 90: 1291–1305.

[55] Johnvesly, B. and Naik, G. R. (2001). Studies on production of thermostable alkaline protease from thermophilic and alkaliphilic *Bacillus* sp. JB-99 in a chemically defined medium. *Process Biochemistry*, 37: 139–144.

[56] Hartley, B. S. (1960). Proteolytic enzymes. *Annual Review of Biochemistry*, 29: 45–72.

[57] Barrett, A. J., Rawlings, N.D. and Woessner, J.F. (1998). Handbook of Proteolytic Enzymes. London, Academic Press.

[58] Tipton, K. F. (1994). Nomenclature Committee of the International Union of Biochemistry and Molecular Biology (NC-IUBMB). Enzyme nomenclature. Recommendations 1992. Supplement: corrections and additions. *European Journal of Biochemistry*, 223(1): 1–5.

[59] Barrett, A. J. (1995). Nomenclature Committee of the International Union of Biochemistry and Molecular Biology (NC-IUBMB). Enzyme nomenclature. Recommendations 1992. Supplement 2: corrections and additions (1994). *European Journal of Biochemistry*, 232(1): 1–6.

[60] Barrett, A. J. and Rawlings, N. D. (1991). Types and families of endopeptidases. *Biochemical Society Transactions*, 19(3): 707–715.

[61] Rawlings, N. D. and Barrett, A. J. (1995). Evolutionary families of metallopeptidases. *Methods in Enzymology*, 248: 183–228.

[62] Knight, C. G., Dando, P. M. and Barrett, A. J. (1995). Thimet oligopeptidase specificity: evidence of preferential cleavage near the C-terminus and product inhibition from kinetic analysis of peptide hydrolysis. *Biochemical Journal*, 308(Pt 1): 145–150.

[63] Hasegawa, J. (1960). Exopeptidases of the human skin. An application of ultramicrotitration and volume measurement methods to a quantitative study of the exopeptidases in skin sections. *Archives of Dermatology*, 82: 595–604.

[64] Parsons, M. E. and Pennington, R. J. (1976). Separation of rat muscle aminopeptidases. *Biochemical Journal*, 155(2): 375–381.

[65] Doebber, T. W., Divor, A. R. and Ellis, S. (1978). Identification of a tripeptidyl aminopeptidase in the anterior pituitary gland: effect on the chemical and biological properties of rat and bovine growth hormones. *Endocrinology*, 103(5): 1794–804.

[66] Rawlings, N. D. and Barrett, A. J. (1997). Structure of membrane glutamate carboxypeptidase. *Biochimica et Biophysica Acta*, 1339(2): 247–252.

[67] Cushman, D. W. and Cheung, H. S. (1971). Spectrophotometric assay and properties of the angiotensin-converting enzyme of rabbit lung. *Biochemical Pharmacology*, 20(7): 1637–1648.

[68] Rawlings, N. D. and Barrett, A. J. (1993). Evolutionary families of peptidases. *Biochemical Journal*, 290: 205–218.

[69] Rawlings, N. D. and Barrett, A. J. (1999). MEROPS: the peptidase database. *Nucleic Acids Research*, 27(1): 325–331.

[70] Rawlings, N. D., O'Brien, E. and Barrett, A. J. (2002). MEROPS: the protease database. *Nucleic Acids Research*, 30(1): 343–346.

[71] Rawlings, N. D., Tolle, D. P. and Barrett, A. J. (2004). MEROPS: the peptidase database. *Nucleic Acids Research*, 32: D160–D164.

[72] Barrett, A. J., Rawlings, N. D. and O'Brien, E. (2001). The MEROPS database as a protease information system. *Journal of Structural Biology*, 134(2–3): 95–102.

In: Proteases: Functions, Mechanisms and Uses ISBN: 978-1-53615-854-0
Editor: Nelson Pérez Guerra

Chapter 2

INSECT DIGESTIVE PEPTIDASES: CURRENT STATE AND PERSPECTIVES

Bianca Santos Henriques and Fernando Ariel Genta*
Laboratory of Insect Biochemistry and Physiology,
Oswaldo Cruz Institute, Rio de Janeiro, Brazil

ABSTRACT

Peptidases are hydrolases, enzymes that are able of cleaving peptide bonds between amino acids residues. These enzymes can be classified according to the position of the bond they hydrolyze (i.e., endo or exopeptidases), catalytic site (i.e., aspartic or cysteine proteinases), or their mode of action. They may also be classified based on their primary structure, into clans and families within major classes (i.e., Family A1 or C1). Proteins are one of the most important components of insect food intake. Therefore, peptidases are fundamental in the breakdown processes of food macromolecules into smaller molecules, allowing absorption of nutrients during digestion. Peptidase enzymology and the physiological understanding of such proteins are significant fields of study due to the financial and health bearings in which many insect species are involved, acting as pests for crops or vectors for causative agents of diseases. In this

* Corresponding Author's E-mail: genta@ioc.fiocruz.br.

way, peptidases are essential targets for the development of inhibitors focusing on the prevention of herbivory, blood digestion in hematophagous insects or even in the invertebrate host-parasite interactions. Four major peptidase groups accomplish insect protein digestion. Serine proteinases, such as trypsins and chymotrypsins, are common to most herbivore and phytophagous species as well as the majority of the hematophagous insects; cysteine and aspartic proteinases, generally found in association, such as cathepsins D and L-like enzymes in hemipterans; and metallopeptidases, as is the case of several aminopeptidases. In this chapter, we will comment on the most recent aspects in the study of protein digestion in insects, detailing particularities of larval and adult main enzymes, adaptations for specific food digestion, and evolutionary tendencies.

Keywords: insect, peptidase, digestion

1. INTRODUCTION

Proteins are major components of insects' nutrition. These macromolecules are composed of chains of consecutive amino acids connected by amide bonds, called peptide bonds. The bonds usually occur between the carbon C-1 on the carboxyl group of one amino acid and the amino-terminal nitrogen N-2 of another (IUPAC-IUB Joint Commission on Biochemical Nomenclature (JCBN) 1984). Peptidases, also known as peptidases or proteinases, are enzymes, which can hydrolyze one or more peptide bonds inside a protein chain. Therefore, insects rely on various peptidases for the breakdown of such large molecules into smaller and absorbable fragments to perform their digestion, allowing amino acids intake by the cells and gut tissue.

Proteolytic enzymes frequently have different specificities in substrate recognition and interaction. According to the model proposed by (Schechter and Berger 1967), the active site of peptidases can be divided into subsites that are numbered outward from the catalytic site, e.g., S1, S2, S3, towards the N-terminus, or S1', S2', S3', towards the C-terminus of the cleavage site in the substrate. Meanwhile, the amino acid residues from the hydrolyzed protein chain (i.e., substrate) are classified as P1, P2, P3 or P1', P2', P3', according to the enzyme subsite in which it is accommodated (Figure 1). The catalytic site of the peptidase hydrolyzes the scissile bond, which is the peptide bond located between P1 and P1' (Figure 1). Although specificity for most peptidases tends to depend on the amino acid residue located in the P1 position, the amino acid sequence flanking the scissile bond also influences the interaction between active site and substrate (Kanost and Clem 2012).

Insects express an abundant number of peptidases with varied roles in the organism. This chapter focuses on the major digestive peptidase classes, gathering the most important aspects and the latest data since the last thorough review (Terra and Ferreira 2012).

2. Classification of Peptidases

There are different criteria for the classification of peptidases. For instance, they can be grouped depending on the position of the scissile bond cleaved amongst the amino acid sequence of the protein. Exopeptidases cleave peptide bonds on the terminal portions of the protein, releasing a single amino acid residue or dipeptides from the amino (i.e., aminopeptidases) or carboxyl terminus ends of the sequence (i.e., carboxypeptidases) (Figure 2A). On the other hand, endopeptidases hydrolyze peptide bonds from non-terminal regions; producing smaller oligopeptides (Figure 2B). There are also oligopeptidases and dipeptidases, responsible for breaking down peptide sequences or pairs of connected amino acids, respectively.

Another system of classification divides the peptidases into subclasses of similar catalytic types. Hydrolases are classified relying on the catalytic mechanism utilized. Thus, the nature of the amino acid residues present in the catalytic site of the enzymes or the cofactor involved in the reaction are determinant (Hartley 1960). According to this method, the so far recognized catalytic types are serine, aspartic, cysteine, metallo, glutamic, threonine, and, as yet, unclassified (Rawlings and Salvesen 2013). Both positional and catalytic classifications fall under the numbered nomenclature for enzymes recommended by the Nomenclature Committee of the International Union of Biochemistry and Molecular Biology (Enzyme Commission). As specified by the commission, peptide hydrolases are named EC 3.4 (http://www.chem.qmul.ac.uk/ iubmb/ enzyme/EC3/4/) (Terra and Ferreira 2005).

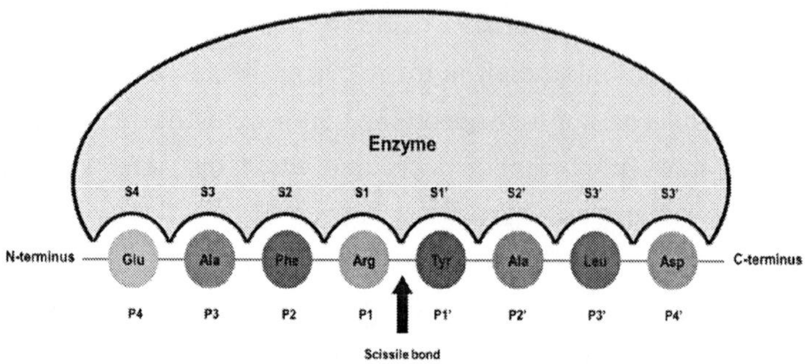

Figure 1. Graphic representation of the Schechter and Berger (1967) nomenclature for binding of a peptide substrate to a peptidase. The peptidase is represented as the enzyme; S1, S1', etc. are subsites for amino acid binding; P1, P1', etc. are the amino acid side chains surrounding the peptide bond to be cleaved, which is indicated by the arrow as the scissile bond.

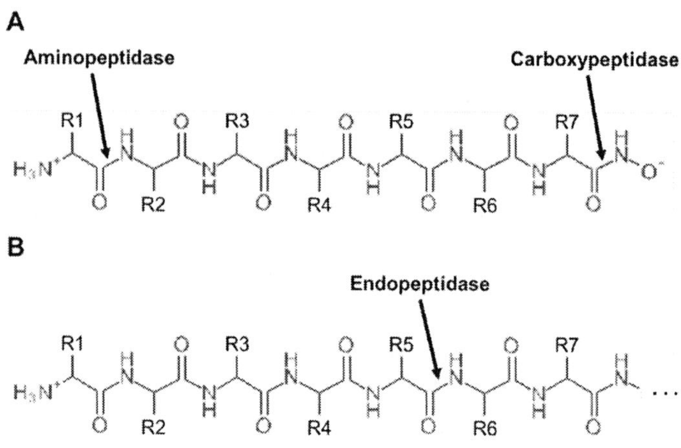

Figure 2. Digestion of proteins. Arrows point to bonds cleaved by respective peptidase type according to site specificity; R, different amino acid moieties. A, Exopeptidase activity on terminal ends, releasing a single amino acid residue. B, Endopeptidase hydrolysis of a non-terminal peptide bond.

Lastly, with the advances in proteolytic enzymes studies came the availability of peptidase amino acid sequences, as well as their tertiary structures. In the 1990s this rising data was used towards the development of an online database based on residues located in the catalytic portion, sequence homology, and evidence of common ancestry (http://www.merops.ac.uk). Peptidases were then divided into Families starting with a "type example" (i.e., archetype), followed by the addition of other proteolytic enzymes with statistically significant similarities. Families can be subdivided into Subfamilies when there is a great difference in primary structure amongst Family members, or even grouped into Clans, following multiple complementary parameters (Rawlings and Barrett 1993, 1999; A. J. Barrett, Rawlings, and O'Brien 2001). Although the MEROPS database is richly integrative and frequently updated, providing sequences alignment, homologs, and reference literature, for example, information can also be found in the hard copy as the Handbook of Proteolytic Enzymes (Rawlings and Salvesen 2013). Further biochemical detailing can be found on the BRENDA Enzyme Database (https://www.brenda-enzymes.org/), which provides, for example, optimum pH values, substrates commonly used, and inhibitors.

3. General and Evolutionary Aspects

Insects are the most diverse group regarding the number of species amongst animals, and were one of the first to colonize successfully terrestrial and freshwater habitats as a direct consequence to the evolution of flight (Misof et al. 2014). This variety infers in a great number of differential aspects, including both physiological and morphological. Despite the huge variability among the insect orders or even families, a general scheme of the insect digestive tract can be drawn as seen in Figure 3. The foregut is a portion that is lined by cuticle, and, includes the pharynx, esophagus, crop, and proventriculus, starting at the oral cavity of the mouth. The crop, depending on the insect order, can be a storage organ or a digestive compartment, whereas the proventriculus can be a triturating organ or a controlling valve, which limits the passage of food to the next portion of the tract. The midgut is the main portion for digestion and nutrient absorption and is composed by a tube, i.e., ventriculus, from which the gastric caecum may branch in some species. The midgut is usually lined by the peritrophic membrane that separates the endo from the ectoperitrophic spaces, compartmentalizing the luminal content. Malpighian tubules are excretory organs that branch from the anterior end of the hindgut. The hindgut is also lined by cuticle, is composed by the ileum, colon, rectum ending in the anus, and is involved in water and ions absorption (Terra 1988).

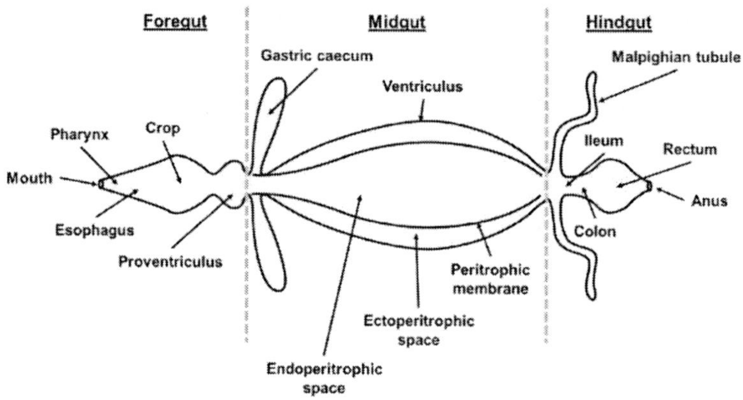

Figure 3. Graphic representation of insect gut compartments (Terra 1988).

Regarding digestive tract morphology and enzymes distribution, initially, it was suggested that these aspects were a direct consequence of the composition of the insect diet (Wigglesworth 1953). Therefore, it was inferred that the kinds of enzymes that could be found in the gut depended upon the adaptation to the habitat and feeding habit of the insect. For instance, sap-feeders, with highly sugary diet, should have little or none digestive peptidases, as in adult Lepidoptera, while predatory insects should display the opposite scenario for the digestion of the large amount of protein ingested, as in *Coleoptera* (Wigglesworth 1953). However, major discrepancies contradicting this theory include, for example, the adult from the black fly species *Cnephia docotensis* (Diptera: Simuliidae), which is a sap-feeder that expresses high levels of peptidase activity in the midgut lumen (Yang and Davies 1968). Meanwhile, phytophagous insects, such as caterpillars or most grasshoppers, are unable to digest cellulose, a major component of leaves, and tend to rely on enzymes from microbiota present in the midgut to break down this macromolecule (Santos et al. 1983; Martin 1983).

Similar discrepancies are found when comparing hematophagous insects from distinct Orders due to different evolutionary paths. This is the case of the mosquitoes (Diptera: Culicidae) and triatomines (Hemiptera: Reduviidae). In mosquitoes, the ingested blood is stored in the posterior portion of the midgut, where it is later digested by serine peptidases like trypsin (Gooding et al. 1973; Schneider et al. 1986). However, in triatomines, the blood suffers a process of concentration in the anterior midgut, where water is absorbed, and erythrocytes are lysed (Barrett 1982; Azambuja et al. 1983). Although recent transcriptome evidence shows that there are large amounts of peptidase transcripts in the anterior portion (Ribeiro et al. 2014), the highest measurable proteolytic activity is found in the posterior midgut of these insects. In this compartment, lysosomal-like peptidases, i.e., cysteine peptidase cathepsin L, and aspartic peptidase cathepsin D, carry out protein digestion (Terra et al. 1988).

In this light, midgut morphology, enzyme distribution, as well as digestion patterns and compartmentalization, show great correlation amongst species with closer phylogenetic relationships. Insects within the same taxon with different feeding habits, but with similar digestive adaptations, may have evolved from common ancestry. Interestingly, insects from different taxon displaying similarities in feeding habits tend to evince distinct patterns of enzymes distribution in addition to structural divergence. Thus pointing to a convergence in niche through alternative adaptations and evolutionary history (Terra 1988, 1990; Terra and Ferreira 1994).

Peptidase activity, as in other enzymes, is influenced by environmental properties, such as the pH. Inside the insect midgut lumen, the pH value is usually between 6 and 7.5, yet it may vary along the digestive tract due to the advent of compartmentalization, resulting in heterogeneous buffering. Exceptions can be found among Lepidoptera, scarab beetles, and nematoceran Diptera larvae, displaying alkaline pH values ranging from 9 to 12. In contrast, cyclorrhaphous Diptera, and heteropteran Hemiptera feature an acidic portion of the midgut, in the 3.1-3.4 pH range (Terra and Ferreira 1994). Although lumen pH may result from ancestral adaptation to specific feeding habit, evolutionary descendants can diverge in habit while maintaining buffering characteristics. Therefore, pH is not necessarily directly connected to the diet, instead, correlating to phylogeny (Terra and Ferreira 2012).

4. Digestion of Proteins

The process of protein digestion starts with the increase in activity of peptidases, initially endopeptidases, which hydrolyze the internal peptide bonds of the chain. Due to the varied amino acid composition of these macromolecules, as well as the relative specificity of substrate recognition and binding, it is necessary for insects to express a combination of different proteolytic enzymes. This allows for wider nutrient exploitation. Oligopeptides resulting from this step of the digestion are then lysed by aminopeptidases and carboxypeptidases, respectively from the amino and carboxyl-terminal ends, releasing single amino acid residues in the process. Finally, dipeptidases hydrolyze any remaining dipeptides. In insects, there are four major classes of digestive peptidases, varying in the mode of action, active site, and even in specificity within each class, i.e., serine, aspartic, cysteine and metallo peptidases.

5. Serine Peptidases

Serine peptidases (EC 3.4.21) are usually characterized by the conserved presence of three amino acid residues in their active site. This catalytic triad is composed of serine, histidine and aspartic acid residues (His 57, Asp 102, and Ser 195, chymotrypsinogen numbering convention), in which the serine acts as the nucleophile, responsible for attacking the peptide bond, and the histidine is a proton donor (Rawlings and Salvesen 2013). According to the

MEROPS classification system, this class of peptidases comprehends 50 Families, distributed into 16 different Clans. Family S1 (Clan PA) is the largest amongst peptidase Families and includes the best known proteolytic enzymes: chymotrypsins and trypsins, the latter being the archetype for the Family. The tertiary structure for these peptidases are composed mainly of β-sheet, and the catalytic units display a two-domain structure, each with a β-barrel, separated by the catalytic cleft. Generally, serine peptidases are synthesized as a proenzyme that requires proteolytic activation. (Rawlings and Salvesen 2013).

6. Chymotrypsin

Chymotrypsins (EC 3.4.21.1) were first enzymes described as being responsible for the second most important proteolytic activity in pancreatic juice from cattle (Rawlings and Salvesen 2013). This enzyme is characterized for the preference for cleavage on the carboxyl side of aromatic amino acids, and, in insects, display an alkaline pH optima ranging from 8 to 11, in addition to the instability at lower pH values (Terra and Ferreira 1994).

This serine peptidase hydrolyzes amide and ester substrates through a three-step process, characterized by the formation of an enzyme-substrate complex followed by an acyl-enzyme intermediate (Rawlings and Salvesen 2013). Insect chymotrypsins bind preferentially to Phenylalanine at the P1 position and were found to be almost inactive when Tyrosine is located at that sub-site. Moreover, they have extended active sites that require longer substrates for successful detection and activation. Therefore, an example of a usual substrate for *in vitro* activity assays is Succinyl-Ala-Ala-Pro-Phe-p-nitroanilide in spectrometry determination, or Succinyl-Ala-Ala-Pro-Phe-HNMec for more sensible fluorescence determination (Terra and Ferreira 2012; Rawlings and Salvesen 2013). Both examples display a Proline at P2, a feature that is common amongst synthetic substrate for chymotrypsins.

Various general serine peptidase inhibitors can inhibit chymotrypsin activity, such as Di-isopropyl fluorophosphates, also known as DFP, which effectively generates an irreversible inhibition of the enzyme a concentration of 0.1 mM. Another commonly used inhibitor is Phenylmethylsulphonyl fluoride (PMSF), also an irreversible inhibitor in a range of 0.1-1 mM, as well as Leupeptin, that competitively inhibits chymotrypsins and can be used in a range of 10-100 μM. Few naturally occurring proteins have also been described as inhibitors for chymotrypsins, e.g., serpins and aprotinin, and synthetic peptides can also be used, like Ile-Val-Asn-Gly-Glu-Glu-Ala-Val-Pro-Gly-Ser-Trp-Pro-Trp, a highly potent competitive inhibitor (Hooper 2002; Rawlings and Salvesen 2013).

Although the resolution of the 3D structure of the fire-ant (*Solenopsis invicta*) digestive chymotrypsin revealed a strikingly high level of similarity with mammalian chymotrypsin, including a shared conserved backbone, comparisons also show great divergence. Sequence information show variation at several loci, thus indicating that both enzymes have distinct modes of activation (Botos et al. 2000). This variation may correlate to the adaptation to different substrates. Chymotrypsins reportedly show distinct levels of expression depending on the plant host the insect feeds on (Ge et al. 2012). Besides, these amino acid residues substitutions generally occur within regions that would be in contact with peptidase inhibitors, inferring an adaptation induced by the frequent interaction with these molecules (Bown et al. 1997).

Furthermore, insect digestive chymotrypsin appears to differ also in biochemical characteristics. For instance, mammalian chymotrypsin has a pH optimum around 8, with two catalytic pKas of 6.8 and 9.5. However, lepidopteran (*Manduca sexta*) chymotrypsin has maximum activity at pH ranging between 10.5 and 11, while showing only one kinetically significant pKa at pH 9.2, a change that is not unprecedented (Peterson et al. 1995). Analyses suggest that this pKa elevation results in changes near the catalytic site, modulating the resistance to modification by chymotrypsin specific plant peptidase inhibitors, secondary metabolite ketones that are similar in action to tosyl phenylalanyl chloromethyl ketone (TPCK), (Tamaki and Terra 2015).

Investigations for the specific mechanism for the insensitivity to ketone inhibitors showed a Phenylalanine substitution with Leucine at position 37 in the chymotrypsin. This change results in loss of important contact sites within the interaction with the ketone, especially in polyphagous insects that often come in contact with ketone-generating plants (Terra and Ferreira 2012). Moreover, gene expression levels can be regulated to overcome the effects caused by the ingestion of such plants through the upregulation and overexpression of chymotrypsin genes (Kuwar et al. 2015; Souza et al. 2016).

Phylogenetic comparisons among chymotrypsins within different insect orders show that lepidopteran digestive chymotrypsins have diverged greatly and form a similarity cluster, separate from Coleoptera and Diptera enzymes, which show similarities to each other. Thence, Lepidoptera have seemingly evolved a unique specialized selection of enzymes, evidenced by the distinction of substrate binding site and activation motifs (Dias et al. 2015). Within Diptera, phylogenetic reconstruction of *Drosophila melanogaster* and *Musca domestica* chymotrypsin sequences show significant similarity evincing expansion of the enzyme genes ancestral to the evolutionary separation between the two species, the latter with 44 sequences found in the transcriptome (Tamaki et al. 2012). Interestingly, in *Phlebotomus papatasi* (Diptera: Psychodidae), the transcriptome analyses resulted in the description of nine novel chymotrypsins expressed preferentially during larval stages, and two with a higher expression after the ingestion of blood (Abrudan et al. 2013). This is a case of adaptation to different types of diet in which the change is triggered by the metamorphosis of the holometabolous insect.

In contrast, Hemiptera, lack this type of enzymes in their digestive tract either in hematophagous species (Balczun et al. 2012; Ribeiro et al. 2014) or in predators (Fialho et al. 2012). However, predator and phytophagous hemipterans express digestive chymotrypsins (along with other proteolytic enzymes) in their salivary gland secretions (Hori 1970; Takanona and Hori 1974; Cohen 1993; Colebatch et al. 2002; Zeng et al. 2002; Zibaee et al. 2012; Mehrabadi et al. 2014). This is said to be due to a role in the extra-oral digestion of the food. In spite of the common presence of these chymotrypsins in the saliva, they tend to appear in minor amounts. In *Creontiades dilutes* (Hemiptera: Miridae) the main peptidase activity in salivary secretions is assigned to chymotrypsins (Colebatch et al. 2001). Authors suggest that this discrepancy owes to a synergetic activity of the proteolytic enzymes, in which small amounts of trypsins are required for the activation of the mature form of chymotrypsin that becomes predominant.

7. Trypsin

Trypsins (EC 3.4.21.4) were initially described in the XIX century as the main proteolytic activity in pancreatic secretions, distinguished from pepsin for their higher optimum pH (Rawlings and Salvesen 2013). In insects, these enzymes are characterized for the pI values between 4–5, and their higher proteolytic activity that can be achieved in pH ranging from 8 to 10, independently of the environmental pH of the insect midgut (Terra and Ferreira 2012). Trypsins preferentially cleave amide substrates on the carboxyl side of basic amino acid residues, like Arginine and Lysine, at P1. Thereby, the catalytic efficiency (kcat/KM) reflects the preference for basic side chains, resulting in a, at least, 105 higher ratio compared with other amino acid residues, while the preference for Arg over Lys is 2 to 10-fold (Rawlings and Salvesen 2013). Lepidopterans are exceptions to this usual preference for Arg at the P1 position. Plant peptidase inhibitors frequently have lysine at that position, an adaptation to avoid hydrolysis by trypsin-like enzymes. Thus, Lepidoptera insects have adapted and prefer Lys at P1 and have more hydrophobic substrate binding sub-sites when compared to other insects' trypsins (Lopes et al. 2004).

The secondary binding site on either side of the scissile bond have minor relevance, though some residues may decrease enzyme activity depending on specific positions. Although maximum activity for mammalian trypsin requires the presence of moderate amounts of calcium in the reaction buffer, usually around 20 mM of $CaCl_2$ (Rawlings and Salvesen 2013), insect trypsins are not activated or stabilized by calcium ions. Besides, they are frequently unstable at lower pH (Terra and Ferreira 1994).

Trypsins have highly conserved tertiary structures, which shows great resemblance to chymotrypsins. Their primary structure can vary considerably, maintaining fold similarity and position of important residues, such as the catalytic triad in the active site and flanking residues likewise. All trypsins are synthesized as a proenzyme. However, in their active form, they consist of a single peptide chain with two β-barrel domains connected by the catalytic residues (His 57, Asp 102, and Ser 195, chymotrypsinogen numbering convention) (Rawlings and Salvesen 2013). Furthermore, regarding the triad, His 57 is considered to act as a general base, while Asp 102 stabilizes the correct tautomer for His 57 and neutralizes the positive charge during the catalytic reaction (Rawlings and Salvesen 2013). Insect trypsins also carry other conserved characteristics. For instance, each complete sequence presents signal and activation peptides, includes the activation amino acid motif (IVGG) on the N-terminal end, three pairs of cysteine residues responsible for disulfide bonds, and the residue Asp 189, which determines the enzyme specificity (Terra and Ferreira 2012).

Because of this need of activation, there is also need of a tight regulatory system to prevent untimely cleavage of the enzyme precursor, i.e., trypsinogen into its active form. Since trypsinogen is commonly activated by hydrolysis catalyzed by an already active trypsin, this activation control is performed by trypsin inhibitors, such as serpins (Meekins et al. 2017) and inhibitors of the kazal-family. The latter is widespread amidst all Hexapoda orders, and phylogenetic analysis shows that their sequences generally cluster together according to their order (Van Hoef et al. 2013). This evidence suggests that the protection of the digestive system through inhibitors is ancestral and as necessary as it is in mammalian animals.

The substrate for these enzymes form an antiparallel β-sheet with the protein-binding site, and the binding specificity is determined by Asp189 located at the bottom of the S1 pocket (Rawlings and Salvesen 2013). Trypsinolysis in insects is usually measured colorimetrically using benzoyl-arginine p-nitroanilide (B-R-pNA, also known as BApNA) or fluorimetrically using benzoyl arginine 7-amino- 4-methyl coumarin (B-R-MCA) as substrates (Terra and Ferreira 2012). Trypsins are also subject to inhibition by typical serine peptidase inhibitors, such as Phenylmethylsulphonyl fluoride (PMSF, 0.1-1 mM), Di-isopropyl fluorophosphate (DIPF, 0.1 mM), and Dichloroisocoumarin (5-100 μM), that act as irreversible inhibitors. Additionally, other compounds inhibit trypsin with higher specificity, like Amidinophenylmethanesulfonyl fluoride (APMSF, 10-100 μM), which inhibits peptidases with specificity for lysine or arginine substrates, and Tosyl-L-lysine chloromethyl ketone (TLCK), which alkylates the His 46 residue located in the active site of the enzyme, and other naturally occurring protein peptidase inhibitors from plants (Hooper 2002; Rawlings and Salvesen 2013).

Plants produce several defensive compounds in various tissues against pathogens and phytophagous insects, such as the protein peptidase inhibitors for the insect midgut trypsins (Ryan 1990). Since inhibitors may result in developmental issues for the organism (Saadati and Bandani 2011; de Oliveira et al. 2014), thanks to the coevolution of both of these molecules, insects have also adapted to evade this defense system and are able to express various peptidases in response to the exposure to these inhibitors, though the induction mechanism is still unclear. Adaptations include the overexpression of trypsins (Bown et al. 1997; Broadway 1997; Gatehouse et al. 1997), inactivation of the peptidase inhibitors through proteolysis catalyzed by the insect's enzymes (Giri et al. 1998), or by the expression of new peptidases that are insensitive to the inhibitor (Mazumdar-Leighton and Broadway 2001a, 2001b).

Available data suggest that the susceptibility of trypsins to plant peptidase inhibitors is a result of molecular differences that modifies the interaction between the inhibitor molecules and the peptidase active site. Lepidoptera differentiates trypsin gene expression depending on the duration of the exposure to the inhibitor. It has been demonstrated that the acute exposure leads to the activation of a small set of genes, whereas chronic exposure induces upregulation of a larger set of trypsin genes (Souza et al. 2016). The differential trypsins can diverge in susceptibility based on inhibitor size when compared to the constitutive form of the enzyme. In this way, the structurally diverse trypsins expressed may limit the access of larger inhibitor molecules to their active site (Mazumdar-Leighton and Broadway 2001b). Further data infer that lepidopteran digestive trypsin has a more hydrophobic surface than coleopteran ones, which creates the tendency to oligomerize, and this hinders the inhibitor binding (de Oliveira et al. 2013; Tamaki and Terra 2015). Concerning alternative modes of inhibitor evasion, still amongst Lepidoptera, the incubation of the soybean trypsin inhibitor with *Helicoverpa armigera* saliva results in almost complete degradation of the protein, while midgut extracts do not seem to bear the same effectiveness (Upadhyay and Chandrashekar 2012).

Orthoptera also adapted to feeding on host plants producing peptidase inhibitors by expressing different trypsins (along with chymotrypsins) to dodge inhibition effects. Midgut transcripts analysis followed by a phylogenetic reconstruction revealed five putative trypsins and 15 putative chymotrypsins that cluster into three evolutionarily conserved groups of serine peptidases (Spit et al. 2014).

Much like chymotrypsins, trypsins can be found in the digestive tract of most insect species, except some cucujiformian Coleoptera and hemipterans, in which this class of enzymes can only be found in the saliva of heteropterans (Terra and Ferreira 1994; Cohen 1993). As of now, only *Poecilocapsus lineatus* (Hemiptera: Miridae) (Cohen and Wheeler 1998) and *Dysdercus peruvianus* (Hemiptera: Pyrrhocoridae) (C. P. Silva and Terra 1994) seem to be exceptions, lacking salivary proteolytic activity. Amongst heteropterans, trypsins are usually responsible for the majority of salivary proteolytic activity (Cohen 1993) and can be found in phytophagous and predator species (Hori 1970; Takanona and Hori 1974; Zeng et al. 2002; Zibaee et al. 2012; Mehrabadi et al. 2014). In the latter, the insect performs chemical liquefaction entirely inside the prey's body, by pumping a one-way flow of enzymes into the insect (i.e., nonrefluxer), consequently turning the prey's exoskeleton into an extension of the predator's digestive tract (Cohen 1995).

Trypsin is the most studied amongst insect digestive enzymes, not only for the dispersion of this proteolytic enzyme throughout insect orders but mainly for its major economic importance due to the large number of insect species that pose a threat to crops as well as stored grains, especially in Lepidoptera. Trypsins have been investigated and described in Coleoptera (e.g., Gooding and Huang 1969; Cheung and Gooding 1970; Levinsky et al. 1977; Colepicolo-Neto et al. 1986; Ferreira and Terra 1989; Zhu and Baker 1999), Diptera (e.g., Gooding and Rolseth 1976; Thomas et al. 1976; Tong et al. 1981; Terra and Ferreira 1983; Graf and Briegel 1985; Graf et al. 1991; Lemos and Terra 1992), Lepidoptera (e.g., Miller et al. 1974; Ahmad et al. 1980; Johnston et al. 1991; Milne and Kaplan 1993; Peterson et al. 1994), Hymenoptera (e.g., Giebel et al. 1971; Jany et al. 1978; Schumaker et al. 1993), Orthoptera (e.g., Sakal et al. 1989; Christeller et al. 1989), Isoptera (e.g., Sethi et al. 2011; Lima et al. 2014), Blattodea (e.g., Tamaki et al. 2014), and Thysanura (e.g., Zinkler and Polzer 1992).

Apart from its importance in peptidase inhibitors evasion, the differential expression of trypsins also enables phytophagous insects to feed in a bigger diversity of host plants. Based on the predicted structures of differentially expressed trypsins, it was possible to notice that these proteins are unlike each other, showing changes adjacent to interaction sites. In turn, this might reflect divergent binding energies with substrates and inhibitors (Chikate et al. 2013). Preferential substrate recognition also occurs in hematophagous insects; this is evidenced by the differential midgut tryptic activity observed in *Phlebotomus duboscqi* (Diptera: Psychodidae) fed with blood from different animal hosts. Sandflies fed with chicken blood display lower levels of trypsin activity when compared with those fed with rabbit blood (Pruzinova et al. 2013).

Amongst holometabolous insects, Culicidae (Diptera) have an interesting pattern of tryptic activity induced after blood meals that has been described in pathogen vectors *Aedes aegypti*, *Anopheles gambiae*, and *Culex quinquefasciatus* (Felix et al. 1991; Dana et al. 2005; Reid et al. 2015). These insects dispose of a biphasic pattern of peptidase activity. Transcription for an early trypsin, which is an exclusively female enzyme for *A. aegypti* (Noriega et al. 1996), starts just after adult emergence from pupae and the mRNA is stored in the midgut epithelium. Within the first hours after blood ingestion, translation of the early trypsin protein is initiated. During the second phase, there is the transcription and translation of the late trypsin, a process that can be activated by several proteins, like albumin or gamma globulin, for example (Noriega and Wells 1999). It has been speculated that the activation of the late trypsin transcription depended on the early trypsin activity. However, later studies applying RNA interference showed that silencing of the early trypsin does not affect late trypsin transcription (Lu et al. 2006).

Aedes albopictus (Diptera: Culicidae) is characterized by a complex profile of tryptic activity, even in preimaginal stages. In this species, there is seemingly a stage-specific differential expression of trypsin isoforms (Saboia-Vahia et al. 2013). A stage-specific gene has also been found in *A. aegypti* larvae (i.e., AAEL005607). This gene was transcriptionally confirmed to being expressed only in mosquito larval stages, increasing the level of expression throughout larval development up to a 5000-fold compared to blood fed adults (Soares et al. 2011). Interestingly, proteomic analysis has been able to identify serine peptidases expressed even in the pupal stage of this species (Fernandes et al. 2016). Equally curious, sugar-fed Culicidae also expresses active trypsins in their midguts, for reasons yet elusive (Saboia-Vahia et al. 2014; Borges-Veloso et al. 2015; Dias-Lopes et al. 2015).

For disease vectors, the interaction with the pathogen may also influence or be influenced by digestive peptidases, such as trypsins. The parasite-host relationship is widely complex and affects various physiological aspects of both organisms involved (Schaub et al. 2016). Although in some vector-parasite interactions the digestive enzymes do not seem to play a part in the midgut colonization, as is the case of tsetse flies (Diptera: Glossinidae), in which the digestive trypsin levels do not affect the development of *Trypanosoma brucei* ssp. (Schaub, Vogel, and Balczun 2016), other species are significantly modulated. Trypsin activity in the midgut of the infected *A. aegypti* is known to impact the replication of dengue virus (DENV) negatively (Molina-Cruz et al. 2005). However, this virus was shown to induce upregulation of pacifastin (Pac2), a strong trypsin-like inhibitor, both in the salivary glands and in the midgut of the insect (Soares et al. 2018).

Interestingly, the transcriptional modulation timing in the salivary glands coincides with DENV transmission period after it has reached the organ, indicating the importance not only for development but also for transmission to the vertebrate host. In parallel, secreted factors from naturally occurring midgut-associated fungi impair trypsin expression and activity in the midgut of *A. aegypti*. Similarly, this reduction in proteolytic catalysis renders the insect more permissive to the virus infection (Angleró-Rodríguez et al. 2017). Moreover, the endosymbiotic bacterium *Wolbachia pipientis* in the same insect species disrupts normal metabolic processes, turning them into less competent hosts. This bacterium infection alters the rate and duration of blood meal excretion; an indication that the digestion might be affected. The delay in the excretion process, however, is not caused by trypsin activity alterations (Pimenta de Oliveira et al. 2017).

However, not only *A. aegypti* is subject to modulation by parasites. *Plasmodium falciparum* infection of *Anopheles donovani* leads to superexpression of a specific trypsin gene in the midgut epithelial cells at day three post-infection, for example (Sriwichai et al. 2012). *Leishmania* spp. parasites are strongly affected by the digestive trypsin activity in the midgut of sandflies (Diptera: Psychodidae), at the same time, the presence of *L. mexicana* in the midgut of *Lutzomyia longipalpis* induces a significant increase in trypsin activity on the second day after blood intake (Moraes et al. 2018). Although it has been demonstrated that the parasite *L. mexicana* is unable to downregulate tryptic activity (Moraes et al. 2018), the silencing of trypsin genes induced by blood-feeding improves survival rates for this *Leishmania* species (Schaub et al. 2016).

On the other hand, sandflies infected with *L. infantum chagasi* present lower trypsin activity levels, evidence for parasite modulation of proteolytic enzymes. The modulation mechanism may the induced by transcriptional downregulation of genes coding for trypsin, or by directly affecting the enzyme activity through the secretion of inhibitors (Schaub et al. 2016). For instance, *L. donovani* expresses two ecotin-like inhibitor genes for serine peptidase when in the midgut of the sandfly *Phlebotomus argentipis* (Verma et al. 2017). Inhibitors promote parasite survival and have also been appointed with an essential role in the differentiation from amastigotes to promastigotes. *Leishmania infantum*, however, has a distinct modulatory mechanism. The promastigote form of this species interferes indirectly in the production of trypsin via the alteration of intestinal pH, which makes the midgut lumen more acidic (Santos et al. 2014).

Since for many vector-borne pathogens there is no current effective vaccine or cure, neglected disease control aims at stopping either vector development, pathogen development inside the vector's organism or even the transmission itself. Because of the widespread resistance development to commercial insecticides due to inadequate management, alternative methods for insect control became a growing market. Thus, the most successful insect pathogen used for insect control is the bacterium *Bacillus thuringiensis* (Bt) and its toxins, namely Cry proteins (Bravo et al. 2011). The Cry toxins are insecticidal proteins naturally produced by the soil-inhabiting bacterium *B. thuringiensis*. The genes for Cry protein encode for protoxins that require proteolytic hydrolysis to become the active toxin form. Then, active toxin interacts with susceptible insect midgut receptors resulting in pore formation in midgut cell membranes, leading to ionic imbalance and consequent septicemia in the insect (Oppert et al. 1997). These toxins are almost exclusively active against different insect orders and has been thoroughly described in various insect species, including mode of action and efficacy (Bravo et al. 2011). Although effective, Bt is also susceptible to resistance development, in many cases through peptidase expression and activity modulation, which is the case for *A. aegypti*. Comparisons between toxin susceptible and unsusceptible strains show that, amongst other peptidases, trypsin levels are upregulated in the resistant mosquito strain, leading to faster protoxin activation and probably protein degradation (Tetreau et al. 2013).

Interestingly, plant peptidase inhibitors, which are part of the herbivory defense mechanism, are also effective bioinsecticides against hematophagous pathogen vectors. Assessment of the effects of a Kunitz-type trypsin inhibitor (LTI) showed that not only this peptidase inhibitor effectively induces a 70% reduction on trypsinolytic activity in *A. aegypti*, but is also a potential insecticidal molecule. LTI affects eggs viability by reducing egg hatching by 40%, delays larval developmental stage by 25%, and induces a mortality level of around 80% (Almeida Filho et al. 2017). Plant peptidase inhibitors from seeds, as well as chimeras constructed based on two Kunitz-type inhibitors through mutation of *Moringa oleifera* flowers trypsin inhibitor (MoFTI) and boophilin, a thrombin inhibitor, are able to induce similar effects on trypsin activity and altogether larval fitness on this vector (Almeida Filho et al. 2018; Soares et al. 2015).

Arthropods are responsible for the majority of agricultural losses that amount for almost 40% of the total production. Since the usual methods for pest control are too environmentally aggressive and generate resistance for its frequent usage, insect digestive enzymes are an alternative target of interest, aiming at the development of novel biotechnological tools to avoid crop predation, focusing mainly on the inhibitor-enzyme coevolution (Macedo and Freire 2011). For this reason, the digestive trypsin of many insect species of economic importance is receiving growing interest from the scientific community. In this way, it is necessary to deepen the knowledge and, therefore, the biochemical characterization, as well as gene identification of these enzymes, have been subject to various recent publications. Trypsins have even been directly appointed as part of the resistance mechanism against chemical insecticides, as in *Bactrocera dorsalis*, in which there is an upregulation of trypsin genes in larvae fed with the pyrethroid β-Cypermethrin, as a detoxification apparatus (Hou et al. 2014). Characterization vary in degree of deepness and detailing, but include economically relevant species from Diptera (H. Chen et al. 2013; Shen et al. 2013; Y. L. Li et al. 2017), Coleoptera (Sharifi et al. 2012; Wielkopolan et al. 2015; Srp et al. 2016; Zulkifli et al. 2018), and, specially, Lepidoptera (Ge et al. 2012; Yao et al. 2012; Chikate et al. 2013; Mahdavi et al. 2013; W. Silva et al. 2013; Ranjbar et al. 2014; Sellami and Jamoussi 2016; Zhou et al. 2016; Sharifloo et al. 2018; Navaei-Bonab et al. 2018).

Alongside with trypsin characterization, this enzyme is also being studied during its interaction with the plant-produced inhibitors to evaluate the efficacy and potential of the practical application on the field. Although insects have developed useful mechanisms to avoid the inhibition effects, they can still cause negative impacts on nutrient exploitation and growth. For instance, the ingestion of potato type II inhibitor reduces trypsin activity in the midgut by up to 75% throughout 24 hours, caused by enzyme sequestration into complexes, which can successfully impair digestion of *Helicoverpa armigera* (Lepidoptera) (Stevens et al. 2013). The same effect can be achieved with the ingestion of the sweet and chili pepper plant (*Capsicum annuum*) inhibitor. However, insects must continuously ingest moderate amounts of the inhibitor since neonatal stages. High dosages lead to an upregulation of peptidase genes that could result in compensation (Mahajan et al. 2013). Kunitz-type trypsin inhibitor (also known as soybean inhibitor) significantly reduces lepidopteran larval mass gain, as well as larvae survival rates, whereas it extends larval stage by delaying pupation (de Oliveira et al. 2014). For Coleoptera larvae, the Kunitz inhibitors lead to an important rate of trypsinolytic inhibition (over 75%) and high lethality, namely 80% for male and 100% for female beetles, in the lowest tested dose (0.1 mM) (Medel et al. 2015).

The plant-produced trypsin inhibitors have been put to use through the development of transgenic plants that overexpress endogenous or exogenous peptidase inhibitors that confer resistance to predation from insects of various families (Macedo and Freire 2011). However, such an ubiquitous defense mechanism aligned with the insect plasticity has quickly led to selective pressure and, consequently, to resistance phenotypes amongst insects. For example, the previously discussed pre-adaptation of insects to deal with these dietary inhibitors, which can even include a broad-range tolerance in polyphagous insects to unusual inhibitors, pose as a major problematic issue (Gatehouse 2011). Despite the contrary aspects, a plant peptidase inhibitor originally from cowpea (CpTI) has been successfully employed in transgenic cotton lines (namely SGK 321) combined with a synthetic *B. thuringiensis* (Bt) toxin gene (Cry1Ac) (Gatehouse 2011). The combination of both protein-encoding genes broadens the protection of the plant against herbivory while reducing the likelihood of generating resistant insects.

Since *B. thuringiensis* toxins are expressed in transgenic plants as an inactive form, insect defense mechanism is usually peptidase-mediated. Bt toxin resistant moths lack a major trypsin that is involved in protoxin activation in other species (Oppert et al. 1997). Exposure to Bt toxins was shown to modulate trypsin expression by inducing upregulation of one trypsin gene, while downregulation of another (Yao et al. 2012). With the increased selective pressure caused by the usage of transgenic plants expressing Bt toxins, there is also an increase in Bt-resistant insects. Therefore, these transgenic insecticidal techniques must be applied parsimoniously.

Alternate approaches to the control of insect development through trypsin inhibition or inactivation are the current focus due to these fast adapting organisms. Because of the growing number of published genomes and transcriptomes, the molecular manipulation of gene expression is an increasingly approachable tool. RNA interference using double-stranded RNA delivered through a semi-synthetic diet has been shown to be a successful way to inhibit trypsin gene expression. Lepidopteran larvae fed with dsRNA cognate to one trypsin gene had a drastic reduction in expression levels of the gene, which led to a delayed weight gain. However, larvae were able to recover in spite of the continuous treatment. Trypsin levels were compensated by the overexpression of another trypsin isoform gene. Silencing of both genes resulted in persistent inhibition, with no larval growth recovery (Chandra et al. 2017). Likewise, the feeding of oriental fruit fly larvae (Diptera: Tephritidae) with dsRNA resulted in significant reduction in trypsin genes expression levels, which led to growth impairment (Y. L. Li et al. 2017). Digestive trypsins from *Tribolium castaneum* (Coleoptera) are also subject to successful RNAi silencing, this beetle, however, has a gene expression compensation, which mitigates the loss in proteolytic activity (Perkin, Elpidina, and Oppert 2017). Understanding which are the RNAi susceptible genes, the gene expression levels, and the compensatory mechanisms, is key to enabling application, although the technique is still far from being deployed into field enforcement.

Insects from all families can be targeted as a host of parasitoid insects, mainly from the Hymenoptera order. Parasitoid species are known to manipulate their hosts to improve suitability and allow their development, this includes changing food preference (Smilanich et al. 2011), regulate food ingestion and nutrient utilization (Huang et al. 2008; Ammunét et al. 2009), metabolism control (Salvador and Cônsoli 2008), and even physiological changes of processes under hormonal control (i.e., metamorphosis and growth) (Harvey et al. 2010). Thus, parasitized insects can be led to sterility or death and, therefore, parasitoid species have been used as crop pest biological control (Kaiser et al. 2017). Interestingly, endoparasitic wasps cause trypsin activity levels to be reduced in the midgut of Lepidoptera larvae (Shi et al. 2013; Rossi et al. 2014).

Another interesting tool involves eco-friendly latex fabricated gold nanoparticles (AuNPs). The nanoparticles tested against trypsins from different insect species, including the arboviruses vector *A. aegypti* and agricultural pests, displayed high levels of trypsin inhibition, between 66 and 89% depending on the species (Patil et al. 2016). These nanoparticles are ideal for the peptidase inhibition task for their specificity to amino acid residues from the active site and because this bond may generate disorganization of the tertiary structure of the enzyme, which precludes further substrate binding (Patil et al. 2016).

Apart from the economic importance of crop pests, another insect with monetary significance is the silk-producing larvae of the moth *Bombyx mori* (Lepidoptera). The silkworms are vulnerable to infections by different pathogens which use their midgut lumen as an entrance point, like viruses such as *B. mori* nucleopolyhedrovirus (*Bm*NPV). This specific virus infects midgut cell before spreading to other insect tissues and is considered one of the most virulent, with the potential to cause serious impacts to the sericulture industry. Proteomics analysis of midgut content revealed that the infection with *Bm*NPV does not induce significant alterations in the expression levels of trypsin-like enzymes (Hu et al. 2015). However, a trypsin (*Bm*tryp) identified in the digestive tract of this insect demonstrated strong antiviral activity against *Bm*NPV. *Bombyx mori* race susceptible to infection by this virus showed higher levels of expression of this trypsin gene in their naïve state when compared to a resistant race. Nevertheless, in the resistant race, infection leads to augmentation of the transcripts levels in the midgut, whereas in the susceptible race the expression rates diminish, indicating the importance of this gene to the insect defense against the virus. Furthermore, treatment of infected larvae with purified *Bm*tryp led to insect normal development, while non-treated larvae had a 100% mortality (Ponnuvel et al. 2012). Thus, this digestive trypsin might become a potential tool in the control of viral infection of *B. mori*.

8. Elastase

Elastases (EC 3.4.21.36)-like are serine peptidases of the chymotrypsin family, whose catalytic site is composed of the triad of hydrogen-bonded amino acid residues (Asp 102, His 57 and Ser 195, chymotrypsinogen numbering convention) (Rawlings and Salvesen 2013). Elastase was firstly reported as a minor proteolytic activity in the midgut of coleopteran larvae, *Costelytra zealandica* (Coleoptera) (Christeller et al. 1989), followed briefly by the description of elastase as a part of the digestive arsenal of the black field cricket, *Telleogryllus commodus* (Orthoptera) for the digestion of proteins (Christeller et al. 1990). Since its original identification in insects, elastases have been further described in numerous other species (Terra and Ferreira 1994).

Insect elastase activity is often identified using chromogenic substrate Succinyl-Ala-Ala-Phe-Leu-p-nitroanilide or Succinyl-Ala-Ala-Ala-p-nitroanilide, as well as through the observation for the inactivity against N-benzoyl-L-Tyr-p-nitroanilide, i.e., B-Y-pNA, a substrate mainly for serine carboxypeptidases, or against N-Benzoyl-L-tyrosine ethyl ester, also known as B-Y-ee, a substrate for chymotrypsin, resistant to trypsin. Another critical characteristic is the resistance to inhibition by TPCK, a chymotrypsin inhibitor (Terra and Ferreira 2012). Elastase is inhibited by general serine peptidases inhibitors, like Phenylmethylsulphonylfluoride (PMSF), trypsin inhibitors from soybean (SBTI), and protein elastase inhibitors (Terra and Ferreira 1994).

Since the main substrate for this peptidase assessment can also be hydrolyzed by chymotrypsins, and these enzymes are frequently resistant to TPCK, due to feeding adaptation as previously discussed, it is not uncommon to misinterpret chymotrypsin activity for elastase, or vice-versa (Terra and Ferreira 2012). However, elastase has been successfully purified and identified in the midguts of the gypsy moth (*Lymantriu dispar*, Lepidoptera) (Valaitis 1995) and the fire ant (*Solenopsis invicta*, Hymenoptera) (Whitworth et al.1998). In the farther, elastase constituted about 6% of the total protein content of the fifth-instar larvae midgut, whereas, in the latter, the substrate specificity of the purified products indicates the presence of two elastase-like peptidases in the digestive tract of the insect. Further analysis of the elastase from fire ants midguts revealed a high level of conservancy with moderate levels of similarities amongst the ant enzyme, oriental hornet chymotrypsin, fruit fly and yellow fever mosquito trypsin, and human leukocyte elastase, maintaining the catalytic site triad homology (Whitworth et al. 1999).

In the tomato moth *Lacanobia oleracea* (Lepidoptera) the elastase activity accounts for about 20% of the total proteolytic activity in the insect midgut (A. M. R. Gatehouse et al. 1999). This activity was measured using specific elastase synthetic substrate (suc-AAA-pNA) and confirmed with elastase-specific inhibitor, elastinal, at low concentrations (< 1 μg/mL) in which this inhibitor is not thought to cause cross-inhibition of other enzymes. Investigation on the effect of plant peptidase inhibitors has different effects on *L. oleracea* elastase. While the cowpea trypsin inhibitor (CpTI) did not substantially affect elastase activity, the soybean Kunitz-type trypsin inhibitor (SBKI) was effective against this peptidase in the moth larvae midgut (A. M. R. Gatehouse et al. 1999).

Kunitz-type peptidase inhibitors from the *Acacia* spp. can often be found in the food bodies of the plant, which are nutritionally valuable rewards produced by plants for their mutualistic ant defenders. The extracted inhibitor was tested against elastases of potential exploiters Coleoptera species *Prostephanus truncatus* and *Zabrotes subfasciatus*, and the mutualistic ant larvae (*Pseudomyrmex gracilis*: Hymenoptera), as the legitimate herbivorous consumer of the plant, to test the hypothesis of the enzyme-inhibitor adaptation. Although the elastases were found to have higher levels of activity in the coleopterans, the ant midgut elastase, along with chymotrypsin, are responsible for the majority of serine peptidase activity and show insensitivity to the plant inhibitor (Orona-Tamayo et al. 2013). Therefore, in spite of not being a direct proof of co-evolution, the evidence suggests an interaction between preadapted mutualistic species.

As a target for insect control, mutagenetically generated inhibitors for serine peptidases induce low larvicidal effects in *A. aegypti*, since elastase accounts for only a minor proteolytic activity in the midgut of this arboviruses vector (Soares et al. 2015). Although the activity is detectable, it corresponds to a 32 times lower activity than the trypsin-generated (Soares et al. 2013). This relatively low activity is the case of most insects. For instance, in the generalist feeder moth *Helicoverpa armigera* (Lepidoptera) the characterization of serine peptidases found in the fed larvae returned only two highly different genes for digestive elastases, compared to 18 trypsin genes and 14 for chymotrypsins (L. N. Gatehouse et al. 1997).

9. Cysteine Peptidases

Cysteine peptidases (EC 3.4.22) are proteolytic enzymes in which the nucleophile that attacks the scissile bond is the sulfhydryl group of a cysteine residue. The catalytic mechanism for these peptidases requires a histidine residue as a proton donor or general base, besides the nucleophile. (Rawlings and Salvesen 2013). The first described cysteine peptidase was papain, which became the archetype for clan CA. Most peptidases grouped in this clan was done so based on the crystallographic structure of the enzymes that is generally formed of two domains separated by the catalytic cleft. The N-terminal domain consists of α-helices, while the C-terminal domain contains a β-barrel (Rawlings and Salvesen 2013). In addition to the catalytic dyad formed by Cys and His present in the active site, two other residues are also important for clan CA peptidases catalytic activity, namely glutamine that helps in the formation of an electrophilic center that stabilizes the tetrahedral intermediate, and asparagine, which is thought to orientate the imidazolium ring of the catalytic histidine (Rawlings and Salvesen 2013).

In the insect midgut, cysteine peptidase activity was first hypothesized based on the optimal pH of 5.0 for the major proteolytic activity described in *Rhodnius prolixus*, suggesting the presence of a mammalian-like cathepsin (Gooding, 1969). The activity was later assigned to a thiol-activated cathepsin B-like (EC 3.4.22.1) enzyme, which was then purified and biochemically characterized (Houseman 1978; Garcia et al. 1978; Houseman and Downe 1980). The cathepsin B was further described as the major digestive peptidase, and about 85% of the proteolytic activity of *R. prolixus* midgut was attributed to this peptidase at pH 5.5 (Terra, Ferreira, and Garcia 1988). The insect midgut hydrolase described was thought to be a cathepsin B-like enzyme because this was the first cysteine peptidase described in animals. However, later it became clear that cathepsin B-like enzymes are able to act as endopeptidases but are more important for their exopeptidase activity. In this way, they are classified as peptidyl-dipeptidases, with the ability to release dipeptides from the C-terminal end of proteins. This preferential exopeptidase activity owes to the existence of a loop that forms a cap to the active-site cleft, and to a pair of histidine residues that can bind the C-terminal carboxylate of the substrate (Rawlings and Salvesen 2013).

Therefore, insect digestive cysteine peptidase activity was later reassigned to cathepsin L-like enzymes (EC 3.4.22.15), which are true endopeptidases that preferentially cleaves peptide bonds with aromatic residues in P2 and hydrophobic residues in P3 position (Terra and Ferreira 2012; Rawlings and Salvesen 2013). Cathepsin L is usually assayed at pH 5-6, using Nα-Benzoyl-DL-arginine-nitroanilide (BAPNA), Nα-Benzoyl-DL-arginine β-naphthylamide (BANA), casein or hemoglobin as chromogenic substrates, since this peptidase is capable of hydrolyzing nearly all proteins (Terra and Ferreira 2012). Fluorogenic substrate Z-Phe-Arg 7-amido-4-methyl coumarin is also commonly used, although the addition of ethylenediaminetetraacetic acid (EDTA) and dithiothreitol (DTT) is encouraged, due to the non-specificity of this substrate hydrolysis (Rawlings and Salvesen 2013). Additionally, activation with sulfhydryl agents, such as the previously cited DTT, and inhibition with trans-epoxysuccinyl-l-leucyl-amido (4-guanidinobutane) (E-64) are indications for the identification of such enzymes (Terra and Ferreira 2012). E-64 is a common cysteine peptidase irreversible inhibitor that is effective in concentrations in a range of 1 – 10 μM, due to the binding of the trans-epoxysuccinyl group to the active thiol group of the peptidase (Hooper 2002).

Cysteine peptidases have been described in many other Hemiptera Heteroptera species, along with a few aphids from suborder Sternorrhyncha (Cristofoletti et al. 2003; Deraison et al. 2004). Digestive cysteine peptidases can also be found in coleopteran from the infraorder Cucujiformia, instead of, or together with serine peptidases, except for cucujiform cerambycid beetles, which lack this type of enzyme (Johnson and Rabosky 2000). Cathepsins L, however, are lysosomal peptidases originally described in the rat liver, as a part of animal intracellular digestion inside lysosome vesicles (Rawlings and Salvesen 2013). Therefore, the actual localization of the cathepsin L emerges as a question for the determination of the peptidase as a true digestive enzyme being secreted into the midgut lumen (Terra and Ferreira 2012).

In this way, the first strong evidence for the secretion of this peptidase into the midgut lumen resulted from a study based on RT-PCR and specific antibodies for three different cathepsin L enzymes from *Tenebrio molitor* larvae. Data revealed that two of these cathepsin L enzymes, CAL2 and CAL3, are mainly expressed in the midgut tissue, whereas the third (CAL1) was detected in the fat body and other tissues, probably acting as a lysosomal enzyme (Cristofoletti et al. 2005). Immunolocalization showed that CAL2 along with CAL3 could be found inside large apocrine vesicles, as well as in the luminal content of the midgut, indicating their action as truly digestive enzymes (Cristofoletti et al. 2005; Beton et al. 2012). Sugarcane weevil, *Sphenophorus levis*, also allowed the immunolocalization of a cathepsin L enzyme to secretion vesicles from the first ventriculus of their digestive tract (Fonseca et al. 2012). Assumingly, the addressing of the cathepsin L to secretion into the midgut content is achieved through the overexpression of these enzymes, which would generate saturation of specific receptors involved in the sorting and transport machinery of the cells (Damasceno et al. 2017).

The current use of cysteine peptidases in food protein digestion by Cucujiformia beetles seem to be a result of an adaptation to circumvent the inhibition of the ancestral major digestive peptidase caused by the naturally occurring serine peptidase inhibitors largely found in the plant seeds on which these insects feed (Terra and Cristofoletti 1996). Thus, the use of lysosomal enzymes in place of or in addition to serine peptidases is an ancestral trait to the whole taxon that probably resulted from gene expansion events leading to the rise of digestive cathepsin L. This is evidenced by phylogenetic analyses that compared cathepsin L sequences from *T. molitor* with those from other organisms, including insect and human sequences, showing that all *T. molitor* cathepsin L sequences probably derived from an ancestral lysosomal cathepsin L, based on the positioning of the sequences amongst clades (Damasceno et al. 2017). Larval gut transcriptomic sequencing of both *T. molitor* and *Tribolium castaneum* corroborates the hypothesis of gene expansion during adaptation since localization of cysteine peptidases on chromosomes indicates that either they are single gene copies, or tandem gene duplications arranged in clusters on chromosomes. Besides, genes within a cluster also groups to form one clade on a phylogenetic tree (Martynov et al. 2015).

Hemiptera insects seem to have followed a different evolutionary path leading to the use of these enzymes in their digestion, although the immunolocalization of cathepsin L enzymes also indicates their secretory pathway through small vesicles (Fialho et al. 2013). Supposedly, hemipteran ancestors were cell-feeder insects similar to Auchenorrhyncha phloem-feeders. Since most phloem and xylem sap are very poor regarding protein content, sap-feeding insects require practically no enzymatic hydrolysis, except for sucrose catalysis, which would explain the loss of digestive peptidases. Therefore, the adaptation to different food sources derived from the reacquisition of the ability to digest polymers through mobilization of lysosomal peptidases (Terra and Ferreira 2012). As predicted by mitochondrial phylogeny, within Hemiptera, the Heteroptera common ancestor seems to have been a predaceous insect, and the transition to hematophagy probably emerged in two distinct events, once in Reduviidae and another in Cimicidae, which includes triatomines and bedbugs, respectively (H. Li et al. 2017). The hematophagous habit in triatomines was a possible result of the association of the predaceous ancestral of these insects with nest-living vertebrates while seeking for shelter. This association led to the opportunistic feeding with other nest-dwelling arthropods, which gave way to the possibility of blood feeding (Monteiro et al. 2018).

Transcriptomic analysis of different heteropteran Hemiptera species identified distinct numbers of cathepsin L genes preferentially expressed in the insects' gut. For instance, in *Nezara viridula*, the southern green stink bug, it was possible to identify 26 cathepsin L genes (Liu et al. 2018), whereas in *Halyomorpha halys*, the brown marmorated stink bug, out of the 40 transcripts for cathepsin L enzymes, 33 had almost exclusive gut overexpression (Bansal and Michel 2018). Both species are characterized as being phytophagous with extremely generalist behavior, with the potential to feed in various economically important plants. Therefore, the considerably expanded gene repertoire in these species suggests transcriptional plasticity that could be employed in response to the diverse host-plant challenges faced by the insect while feeding in different food sources (Bansal and Michel 2018).

Interestingly, comparisons between *R. prolixus* digestive tract transcriptome library and whole body library revealed a significantly smaller number of cysteine peptidases gene transcripts localized to this organ. The analysis showed only 11 genes that have a larger expression in the digestive tissues, among which, 10 presented the all characteristic active site residues of the family, and three are overexpressed specifically in the posterior midgut of this insect (Ribeiro et al. 2014). Phylogenetic reconstruction of these sequences and cysteine peptidases from various insects grouped the majority of *R. prolixus* genes with those from other Triatominae forming two distinct triatomine subclades (Ribeiro et al. 2014). In an alternative approach, focusing exclusively on the C1 peptidase family, phylogenetic analysis of *R. prolixus* genes and other hemipteran insects with available transcriptomes revealed a different gene distribution throughout the tree clades. *R. prolixus* genes show a scattered distribution, displaying only two remote small branches grouping three and four gene entries, respectively, of the total 17 sequences analyzed (Henriques et al. 2017). This pattern suggests that the C1 peptidase gene duplication probably occurred previously to the diversification of the Triatominae branch. Thus indicating that in spite of the crucial importance of this peptidase family for protein digestion in hemipterans, the high number of gene copies is unlikely to have been developed during the adaptation of these insects to hematophagous habit, being probably a pre-adaptation.

Since many phytophagous species rely on cathepsin L-like enzymes during food digestion, a large number of studies involving cysteine peptidases focus on the description and characterization, including cloning and purification, of peptidases from economically important insects, like crop pests. The pressure of coevolution between plants and herbivore insects led to the development of host strategies to prevent predation from insects with cysteine peptidases as their major proteolytic activity, much like what happened with trypsin-bearing insects (see item 4.1.2). In this way, plants are able to produce proteins, such as phytocystatins that act as inhibitors for cysteine peptidases in the insect's midgut. This feature can even be manipulated through transgenic approaches to enhance defense function towards pests (Martinez et al. 2016). With the purpose of finding new insect control alternatives, amaranth cystatin is one of the many plant-produced cysteine peptidase inhibitors that have been isolated, characterized, and tested against insect digestive peptidases with important effects (Valdés-Rodríguez et al. 2015).

For instance, oryzacystatin I and II are plant cysteine peptidase inhibitors originally isolated from rice that have been successfully used in individually expressing transgenic crops that are used in the control of pests that predominately use cysteine proteinases, such as coleopteran insects. However, the recombinant expression of these inhibitors individually in potato exhibited little or minimal effects against the Colorado potato beetle (*Leptinotarsa decemlineata*) (Cingel et al. 2015). Interestingly, the co-expression of both inhibitors led to results that are more promising. Early larvae fed with transformed leaves exhibit premature molting, which later reflects in lower weight gain accompanied by a reduction in proteolytic activity levels measured in the larval midgut. Unfortunately, the cysteine peptidase activity inhibition decreased with the long-term permanence of the contact with the transformed plant (Cingel et al. 2017). Coleoptera have also been reported to exhibit distinct cysteine peptidase expression levels depending on the host plant in which the insect feeds (Amiri et al. 2016). Furthermore, studies show that the cathepsin L of Colorado potato beetle has population-associated variability profiles that implicate on differential susceptibility to tomato cystatin (SlCYS8) (Rasoolizadeh et al. 2018). This evinces that insects display great plasticity and can quickly adapt when challenged by cysteine peptidase inhibitors, and, therefore, the use of these tools in pest control still needs careful evaluation.

The western corn rootworm (*Diabrotica virgifera*) is a Coleoptera that requires continuous availability of corn. Thus the adoption of an annual rotation system between corn plants and soybean provided a simple solution for the control of these insects (Levine and Oloumi-Sadeghi 1996). Since soybean is a non-host plant that is rich in cysteine peptidase inhibitors, the selective pressure caused by this crop rotation led to the emergence of a rotation-resistant western corn rootworm population (Levine et al. 2002). The evading mechanism involves an approximately three times higher baseline cathepsin L activity in the midgut of fasting insects that remains significantly higher after exposure to the soybean foliage, which allows longer survival rates and, consequently, greater herbivory damage. The higher proteolytic activity is caused apparently by the overexpression of a single cathepsin L gene (DvRS5) in about 25% when compared to the inhibitor susceptible strain (Curzi et al. 2012).

The selective pressure generated by the digestive peptidases of phytophagous insects upon the stress-induced cysteine peptidase inhibitors is known to create a positive selection that contributes to the functional diversification of these molecules (Christeller 2005). This diversification is brought forth by single substitutions to closely related amino acids at positively selected sites (Rasoolizadeh et al. 2016). Digestive cysteine peptidases, on the other hand, display hypervariable sites, which are distributed within, or close to, amino acid regions that are involved with the interaction with cystatins. Feeding *Leptinotarsa decemlineata* (Colorado potato beetle) with positively selected cystatins, with confirmed cysteine peptidases inhibition effects, induces a compensatory adjustment that leads to an alteration in the cathepsin L expression pattern, doubling the total number of peptidases detected in the insect midgut. The profile change restores cathepsin L activity, as well as foliage consumption rates, to control levels in only 72 hours (Vorster et al. 2015).

In addition to naturally occurring or transgenically induced cysteine peptidase inhibitors, cysteine peptidases are also target to differential inhibition or silencing approaches on insect control. For instance, the supplementation of artificial diets with the entomopathogenic *B. thuringiensis* spores dramatically decreases cysteine peptidases activity levels, whereas increasing larval and pupal period and inducing up to 100% mortality, depending on the spore concentration (Malaikozhundan and Vinodhini 2018). Gene knockdown through RNAi silencing techniques is the newest focus on the field. However, effectiveness may vary among insect populations and should be carefully evaluated before wide deployment. Insects fed with artificial food containing double-stranded RNA targeting the cysteine peptidase with the highest expression level in the digestive tract of the western corn rootworm (DvRS5) had significant decreased transcript levels of this cysteine peptidase in the midgut, whereas the reduction of the proteolytic activity occurred only in a range of 50 - 60% (Chu et al. 2014). In *T. castaneum* there is a similar effect in the silencing of the most important digestive cathepsin L gene. This is due to a triggering of a compensatory mechanism in which other cysteine peptidases are up-regulated, along with some serine peptidases, allowing the insect to maintain the digestive fitness (Perkin et al. 2017).

10. Aspartic Peptidases

Aspartic peptidases (EC 3.4.23) are characterized by being active at acidic pHs. They are named for the presence of two aspartate residues in the catalytic site that act as substrate ligands, plus binding and activating the catalytic water molecule. These peptidases differ from most in that all enzymes from this class are endopeptidases whose attack of the scissile bond comes from the activated water molecule acting as a nucleophile (Rawlings and Salvesen 2013).

This class of proteolytic enzymes is further subdivided into clans and families, amongst which is the family A1 from the clan AA. This large peptidase family is grouped according to similarity to human pepsin A, which is the type enzyme. A1 peptidases are bi-lobed molecules formed through gene duplication of one of the lobes, including the homologous Asp residues that form the catalytic dyad, and between the two lobes is located the active site. The active site cleft is centrally located in the peptidase molecule, with the Asp residues surrounded by mostly identical motifs. The substrate binding cleft can accommodate seven or eight amino acid residues, preferentially cleaving between two hydrophobic residues (Rawlings and Salvesen 2013).

Insect aspartic peptidases were first reported in the housefly, *Musca domestica*, yet still referred to as pepsin or pepsin-like activity in larvae whole body homogenates, with high proteolysis levels at pH ranging from 1.5 to 3.5 (Greenberg and Paretsky 1955). Later on, biochemical characterization of this housefly proteolytic activity distinguished it from a pepsin and assigned it to a digestive cathepsin D restricted to the midgut lumen of this insect, based on features like substrate preference and inhibitors responsiveness (F. Lemos and Terra 1991). Evidences suggest that this activity is important for necrophagous flies due to the ingestion of food highly colonized by bacteria. The ingested bacteria are lysed in the midgut combining the action of the low pH, and enzymatic activity of lysozyme and cathepsin D (Espinoza-Fuentes and Terra 1987). Interestingly, digestive cathepsin D of *M. domestica* lacks the polyproline loop, a common characteristic of intracellular cathepsin D. The lack of this motif brings resemblance to vertebrate pepsin and is evidence to the seemingly adaptation of these insects to deal with the high numbers of bacteria ingested with their food through acidic digestion. Therefore, luminal cathepsin D seems to have evolved from the same archetypical gene as the intracellular cathepsin D, as what happened with vertebrates. The absence of the proline loop seems to have a role in the extracellular activity of both digestive cathepsin D and pepsin (Padilha et al. 2009).

Cathepsin D (EC 3.4.23.5) was a term originally introduced to denominate acidic intracellular peptidases, with catalytic mechanism similar to pepsin. Original assessment method was done with denaturated hemoglobin as substrate, at pH around 3.5, quantifying the soluble peptide products in the supernatant after treatment with trichloroacetic acid (Rawlings and Salvesen 2013). However, synthetic substrates are also available for this peptidase, consisting of long amino acid chains. For instance, for chromogenic assays, aspartic peptidase activity can be measured using Pro-Thr-Glu-Phe-Phe(4-NO2)-Arg-Leu as a substrate, in which the Phe(4-NO2) corresponds to p-nitro-L-phenylalanine (Dunn et al. 1986). For fluorogenic assays, internally quenched fluorescent substrates are used (Pimenta et al. 2001). In both cases, the scissile bond is localized between the two phenylalanine residues. For further enzyme characterization, inhibition of cathepsin D can be achieved with the aspartic peptidase inhibitor pepstatin. This inhibitor is highly selective, acts as a reversible peptidomimetic and is effective at a final concentration of 1 μM. Moreover, several inhibitor molecules were developed based on the pepstatin structure (Hooper 2002; Rawlings and Salvesen 2013).

Apart from houseflies, cathepsin D can also be found in several families of the Cucujiformia infraorder (Coleoptera), occurring together with cysteine peptidases in their midgut in addition to, or replacing, serine peptidases (Terra and Ferreira 1994, 2012). Insects from these families usually feed on seeds that are rich serine peptidase inhibitors, and, therefore, this substitution or addition to their proteolytic repertoire indicates an important adaptation to avoid the inhibition effects, a feature that probably was originated in the common ancestor of the taxon (Terra and Ferreira 2012). Since many coleopteran species feed on economically important plants, the majority of studies involving cathepsin D focus on biochemical characterization of larval midgut peptidases of cereal pests, like, for example, *Oulema melanopus* (Wielkopolan et al. 2015).

Plants, however, can display aspartic peptidase inhibitors in their composition as well, although only in a reduced number of species. Thus, one aspartic peptidase inhibitor from the plant *Lupinus bogotensis* with the highest activity on these enzymes was purified and tested against the coffee berry borer (*Hypothenemus hampei*, Coleoptera), a stored coffee bean pest. Larvae reared on inhibitor containing coffee had an elevated rate of mortality, ranging from 30 to 90% depending on the concentration provided (Molina et al. 2014). This preliminary result raises the aspartic peptidase inhibitors as a putative candidate for insect pest control, with the possibility of transgenic transfer of the encoding gene to the coffee plant, as is commonly done with serine peptidase inhibitors (see item 4.1.2).

Similarly, Hemipteran insects also have cathepsin D-like enzymes present in their midgut. Initially reported in the hematophagous insect *Rhodnius prolixus* (Hemiptera: Heteroptera), it was characterized as an acid peptidase with maximum activity at pH 2.8 using hemoglobin as substrate (Houseman and Downe 1982). Then, cathepsin D was also found in several other Hemiptera, including predacious or seed-feeder species, with maximal activity at pH 2.5 or 3.0 (Houseman and Downe 1983). Later reassessment of the aspartic peptidase of *R. prolixus*, confirmed it to be cathepsin D, and assigned it as a minor hydrolytic activity in the midgut these insects, corresponding to only 15% of the total proteolytic activity at pH 5.5 (Terra et al. 1988). This activity could be possibly involved with the hydrolysis of cysteine peptidase inhibitors in phytophagous insects (Pimentel et al. 2017). Cathepsin D is believed to have a similar evolutionary history to cysteine peptidases in Hemiptera (Terra and Ferreira 1994).

In *Dysdercus peruvianus* (Hemiptera: Heteroptera), cathepsin D, also lacking the proline loop, is secreted into the first part of the midgut, where this aspartic activity amounts to about 40% of the total proteolysis. As the food is transferred from the first midgut region to the second, so does part of the cathepsin D enzymes. However, cathepsin L activity is so predominant that it overcomes the minor cathepsin D present in that compartment (Pimentel et al. 2017). In this way, data led to the conjecture that cathepsin D acts hydrolyzing the cysteine peptidase inhibitors ingested during feeding on plant seeds.

The genome sequencing of *R. prolixus* highlighted the finding of lineage-specific cathepsins D occurring in tandem clusters, probably due to the replacement of the digestive serine peptidases (Mesquita et al. 2015). Transcript expression analysis of digestive gene libraries of the same insect, revealed 17 coding sequences for cathepsin D enzymes, that are highly expressed in the anterior midgut. Amongst the appointed sequences, only one presents the proline loop, two appears not to be digestive, and the other 13 form a single clade (Ribeiro et al. 2014). In *Triatoma infestans*, characterization of midgut cathepsin D genes showed that, of the two, one does not have a proline loop (TiCatD) and groups with digestive cathepsin D from other insects and is regulated by feeding, whereas the second has a proline loop (TiCatD2) and identifies with lysosomal cathepsin D (Balczun et al. 2012). Interestingly, phylogenetic reconstruction evinces that both cathepsins D above with this motif groups together in the same clade, probably without digestive functions (Ribeiro et al. 2014).

Meanwhile, analysis based on the genome data of *R. prolixus*, and comparisons with other arthropods' genome datasets, suggested a putative gene expansion deriving from a single clade composed of 15 sequences of family A1 peptidases from *R. prolixus* (Henriques et al. 2017). The high number of similar sequences of these peptidases in the *R. prolixus* genome suggests a recently acquired set of copies, probably after the differentiation of main hemipteran suborders. This led to the hypothesis of the involvement of these genes in the adaptation to hematophagy since most are actively expressed in the midgut and present putative secretory pathway signal peptides, as well as conserved cleavage site residues. Furthermore, the frequency and distribution of basic residues on the protein surface constructed based on homology structural models, shows an exemplary cathepsin D-like structure, reinforcing previous classifications (Henriques et al. 2017).

Regarding the vector-parasite interaction within Triatominae (Hemiptera: Reduviidae), the colonization of *T. infestans* midgut with *Trypanosoma cruzi* upregulates the gene encoding cathepsin D enzymes in this digestive compartment (Buarque et al. 2013). The change in the expression patterns corroborates previous data that showed that infection of *R. prolixus* with *T. cruzi* induces a significant elevation in cathepsin D activity levels in the midgut, in the first days after blood ingestion (Borges et al. 2006).

The medical use of maggots of the common green bottle fly, *Lucilia sericata*, is an approved approach to the treatment of many types of chronic and necrotic wounds. Modern maggot debridement therapy involves the physical removal of the necrotic tissue with the mandibles of the maggot, concomitantly with the release of digestive enzymes into the wound, which act in the pre-oral digestion. The differential transcriptome of larvae of L. sericata, revealed about only 3% of the peptidases found are classified as aspartic peptidases. However, amongst them most are cathepsin D and, with the exception of two, all clusters lack the proline loop, and are predominantly expressed in the midgut (Franta et al. 2016). These characteristics are congruent to those found in *M. domestica*, indicating their important role in digestion and the elimination of bacteria present in the ingested food.

As a target for insect control, digestive cathepsin D silencing through insect feeding with artificial diets containing double-stranded RNA was shown to be an effective tool. Examination of the outcome of increasing concentration of dsRNA revealed a positive correlation with increased mortality of *Diaphorina citri*, one of the most important citrus pests (Galdeano et al. 2017). However, the change in dsRNA concentration did not induce higher levels of gene expression silencing, indicating that the highest doses induced mortality through off-targets effects or caused by saturation of the RNAi pathway. Besides, the incorporation of the dsRNA onto plant leaflets by soaking technics also showed promising results, since dsRNA could be detected in the leaves after 11 days and the insects fed had reduced gene expression levels accompanied with elevated rates of mortality (Galdeano et al. 2017). Therefore, using moderate concentrations of dsRNA, it is possible to use oral target mRNA knockdown to further studies in gene functionality and for the development of novel insect control tools.

11. Aminopeptidases

Aminopeptidases (E.C. 3.4.11) comprise a class of exopeptidases, which according to the MEROPS classification system can be assigned to various families depending on their sequence homology, although most are metallopeptidases due to their dependence on metal ions for activation. Aminopeptidases are capable of sequentially removing N-terminal amino acids from peptides, and are classified based on their dependence on metal ions (generally Zn2+ or Mn2+) and substrate specificity (Terra and Ferreira 2012). In metallopeptidases, the nucleophilic attack on the scissile bond is mediated by a water molecule, which is activated by a divalent metal cation (Rawlings and Salvesen 2013).

Insect midgut metallo aminopeptidases can be found in two different forms. In more basal insect orders, such as Orthoptera, Hemiptera, Coleoptera Adephaga, they are found in major amounts as a soluble hydrolase, whereas in more derived species like Coleoptera Polyphaga, Diptera, and Lepidoptera they are bound to the microvillar membranes of the midgut cells (Terra and Ferreira 1994). In insects, optimal pH ranges between 7.2 and 9.0, with a broad specificity, being able to hydrolyze a variety of amino acyl β-naphthylamides substrates, except acidic amino acyl β-naphthylamides, which indicates that they are mostly aminopeptidases N (Terra and Ferreira 2012). Aminopeptidase N (EC 3.4.11.2, MEROPS family M1) is a broad spectrum hydrolase that was named for its preference for action on neutral amino acids, primarily removing alanine and leucine residues from peptides (Rawlings and Salvesen 2013). The overlapping of substrate specificities and tissue distribution hinders differentiation between aminopeptidase N and leucyl aminopeptidases (E.C. 3.4.11.1, MEROPS family M17), and, therefore, is possible that many digestive enzyme characterization results from confusion of the two (Rawlings and Salvesen 2013). Inhibition can be achieved through the addition of metal chelating agents to the reaction. Ethylenediaminetetraacetic acid (EDTA) is the most commonly used chelator, which reversibly inhibits metallopeptidases at a final concentration of 1–10 mM, followed by 1,10-Phenanthroline, another chelating compound that reversibly inhibits this type of enzymes at 1-10 mM (Hooper 2002). Although inhibition can also be generated by peptides, e.g., amastatin and bestatin, which are competitive and reversible inhibitors that can be used in a range of 1-10 μM (Mucha et al. 2010; Rawlings and Salvesen 2013).

Interestingly, evidences presented to the moment indicate that coleopterans present a single aminopeptidase type in their midguts, as evidenced in *Attagenus megatoma* (Baker and Woo 1981) and *T. molitor* (Cristofoletti and Terra 1999; A. H. P. Ferreira et al. 2007; Zulkifli et al. 2018). This coleopteran N-terminal exopeptidase activity has been assigned to an aminopeptidase N. Diptera and Lepidoptera, however, differ in that they present multiple aminopeptidases varying in substrate specificity, as best described in *Rhynchosciara americana* (Diptera) (Terra and Ferreira

2012). This insect displays an aminopeptidase A that preferentially removes N-terminal aspartic acid or glutamic acid residues from peptides, and three midgut aminopeptidases N (one soluble and two membrane-bound) (Klinkowstrom et al. 1995). More data is still necessary to confirm this hypothesis. Moreover, comparisons between several aminopeptidases N from coleopterans and lepidopterans show that despite the similar genomic organization, both sequences from these orders are phylogenetically distant. In addition, the number of aminopeptidases found in Coleopteran genomes indicates an expansion generated by tandem gene duplication events that probably occurred after the evolutionary divergence of the two groups (Van Munster et al. 2011).

Sequence analysis of the available insect midgut aminopeptidases N, reveals the presence of a signal peptide, a conserved RLP motif near the N-terminal end, a zinc-binding motif HEXXHX18E, a GAMEN conserved motif, and a long hydrophobic C-terminal containing a GPI anchor (Terra and Ferreira 2012). Furthermore, phylogenetic reconstructions group lepidopteran aminopeptidases into at least four major distinct branches of enzymes, which may be due to substrate specificities or to the necessity of peptidase inhibitors evasion during feeding (Terra and Ferreira 2012). Corroborating this hypothesis, analysis of expression levels of various aminopeptidase N genes from *Epiphyas postvittana* (Lepidoptera) shows that transcripts prevalence changes depending on food source (Simpson et al. 2008).

Hemipteran insects seem to have aminopeptidases with similar substrate preference to coleopteran digestive aminopeptidases, as evidenced in the sap-sucking *Acyrthosiphon pisum* (Cristofoletti et al. 2006), in the phytophagous *Dysdercus peruvianus* (Costa et al. 2011), in the predator *Podisus nigrispinus* (Fialho et al. 2012) and in the hematophagous species *R. prolixus* (Houseman and Downe 1981). In the latter, aminopeptidase activity seems to be distributed along the entire midgut, although with higher specific activity in the posterior portion of the compartment and is majorly trapped between the outer and inner microvillar membranes (Terra et al. 1988; C. Ferreira et al. 1988). *R. prolixus* aminopeptidase activity can be detected associated with the membranes of the microvilli, with extracellular membrane layers, with the lysosomes of the midgut cells, as well as in the rough endoplasmic reticulum and at the periphery of intracellular storage vesicles, suggesting a role in the terminal digestion of the blood protein (Billingsley and Downe 1985).

Analysis of the genomic dataset of the same species revealed a putative recent expansion in the leucyl aminopeptidase family M17, possibly divided into two duplication events during the evolution of the Triatominae group. Despite the cytosolic nature of the majority of the sequences characterized as members of this family, many display putative signal peptide sequences, or non-cytosolic and transmembrane topologies (Henriques et al. 2017). Since digestive leucyl aminopeptidases have been already described in other insect species (e.g., Okuda et al. 2005; Božić et al. 2008) it is possible that this expansion is connected with blood protein digestion. However, more studies are needed to clarify the function of these genes.

Midgut aminopeptidase N has been characterized as an important receptor for *B. thuringiensis* toxins (Pigott and Ellar 2007), possibly due to their localization in the microvillar membranes (Terra and Ferreira 2012). Since Cry proteins are subject to major attention focusing on the control of insects, both in crop pests and disease vector species, evidence for the role of this enzyme as a toxin receptor have been described in many insects, such as Lepidoptera (Knight et al. 1995), Coleoptera (Pauchet et al. 2009; Bulushova et al. 2011; Shu et al. 2015; Jiang et al. 2017), and dipteran vectors *A. aegypti* (J. Chen et al. 2013) and *Culex quinquefasciatus* (Rezende et al. 2017). Furthermore, there is evidence proposing that aminopeptidase N isoforms have differential toxin binding properties (Nakanishi et al. 2002; Rajagopal et al. 2003; Crava et al. 2013), being even implicated in *B. thuringiensis* toxin resistance. For instance, the transcriptome of naturally occurring *A. aegypti* toxin-resistant strain shows a reduction of about 40% in aminopeptidase N transcripts (Lee et al. 2014). Similarly, *Ostrinia furnacalis* (Lepidoptera) also displays significant down-regulation of aminopeptidase N transcript levels in laboratory-selected resistant strains of the species (Zhang et al. 2017). On the other hand, RNAi silencing of aminopeptidase N in *Helicoverpa zea* (Lepidoptera) results in Cry1Ac resistant insects, revealing the importance of the peptidase in insect control (Wei et al. 2016).

12. Carboxypeptidases

Carboxypeptidases are enzymes, which can hydrolyze peptide bonds releasing single amino acids from the C-terminal end of the peptide chain. Insect digestive carboxypeptidases can be categorized into two different types of metallocarboxypeptidases from family M14. The first is carboxypeptidase A (EC 3.4.17.1) that prefers substrates with aromatic or branched-chained C-terminal amino acid, the second is carboxypeptidase B (EC 3.4.17.2), which has high specificity for excising C-terminal lysine and arginine residues from peptides and proteins, with a preference for arginine (Terra and Ferreira 2012; Rawlings and Salvesen 2013). Insect carboxypeptidase activities are usually assayed in alkaline pH range using the chromogenic substrates hippuryl β-phenyllactic acid (also known as Z-GF) or hippuryl-l-arginine (Z-GR), for carboxypeptidase A or B, respectively, the former with an optimal pH between 7.5 and 9.0 (Terra and Ferreira 2012).

Carboxypeptidase A is a family of proteins with highly conserved amino acid structure, ranging between about 60 and 80% of sequence identity. Including amino acid residues such as histidine and glutamate, which are putative ligands to the catalytic zinc ion, the cysteine residues that form the disulfide bonds and other amino acids that proposed to be important to specificity and catalysis are also conserved (Rawlings and Salvesen 2013). These enzymes are widely spread among insect orders and have been described in various species (Terra and Ferreira 1994). Their sequences present signal and activation peptides along with other typical characteristics, like the zinc ligand amino acids, residues responsible for substrate binding, as well as Arg 127 and Glu 270 (carboxypeptidase A numbering), which are responsible for the peptide bond hydrolysis (Terra and Ferreira 2012).

Carboxypeptidase B, on the other hand, shows strong similarity with carboxypeptidase A, presenting the conserved zinc-coordinating residues His 69, Glu 72, His 196 (carboxypeptidase A numbering), along with other important residues for catalysis like aspartate, arginine, and glutamate in conserved positions. Differences between enzymes are due to substitutions of catalytically important residues (Rawlings and Salvesen 2013). Carboxypeptidase B is similarly found in the midgut of many insect species (Terra and Ferreira 1994). *Helicoverpa zea* is a lepidopteran insect that is known to evade plant serine peptidase inhibitors present in their food by using inhibitor-insensitive peptidases, including carboxypeptidases B. The resolution of the crystal 3D structure of their carboxypeptidase B revealed that the rearrangement of two stabilizing regions, which causes the displacement of the active site cleft, leading to the impairement of the interaction with the potato carboxypeptidase inhibitor, consequently rendering the enzyme resistant (Bayes et al. 2005).

Aedes aegypti carboxypeptidase A has an expression pattern that resembles that of the late trypsin (see item 4.1.2), with the transcript and protein levels reaching a peak 24 hours after the blood ingestion (Edwards et al. 2000). Regulation, however, follows a distinct pathway. Signaling is dependent on the activation of the Target of Rapamycin kinase (TOR), a conserved nutritional sensor, mediated by both amino acids and insulin (Zhao et al. 2014). Interestingly, in spite of the shared similarities between *A. aegypti* carboxypeptidase A and that of *Anopheles culicifacies*, reaching 72% of sequence identity, the expression pattern in the latter reaches a peak in mRNA abundance in the midgut at three hours after the blood meal (Kumar et al. 2014).

Regarding the digestive physiology during insect-parasite interactions, it has been demonstrated that the infection of Culicidae mosquito, *Anopheles stephensi*, with the malaria etiologic agent, *Plasmodium berghei*, induces a significant up-regulation of the carboxypeptidase A gene, followed by an increase in the respective proteolytic activity in the insect digestive tract. Thus, the addition of carboxypeptidase A-directed antibodies in the blood meal containing the parasite significantly reduced the insect infection rates, while blocking the development of *P. berghei* in the midgut (VenkatRao et al. 2017). Therefore, mosquito carboxypeptidase A is a possible tool in the development of transmission-blocking vaccines as a way to interrupt the malaria parasite transmission.

Carboxypeptidase B1 has a similar role in the *A. stephensi-P. falciparum* interaction, the peptidase expression levels increasing upon parasite gametocyte ingestion, whereas the presence of carboxypeptidase B1-directed antibodies also reduces the infection rates (Raz et al. 2013). Together with previous analogous results on *Anopheles gambiae* (Lavazec et al. 2007), this indicates that carboxypeptidase B could also be used in transmission-blocking vaccines against the two main malaria vectors in Africa (*A. gambiae*) and Asia (*A. stephensi*), as well as in overlapping areas where other sympatric vectors can be found, due to the high similarities among enzymes. Furthermore, carboxypeptidase B is also known to interact with the dengue virus (DENV2) E protein in the endoplasmic reticulum of midgut cells in *A. aegypti*. In this way, overexpression of this peptidase gene leads to an intracellular accumulation of the DENV2 genomic RNA or virus particle, reducing virus release. This led to the hypothesis that the binding of the insect carboxypeptidase B with the virus protein inhibits virus RNA encapsulation, consequently stopping the budding from the endoplasmic reticulum, and possibly interfere with immature virus transportation to the trans-Golgi network (Tham et al. 2014).

Initial reports on *R. prolixus* (Hemiptera) digestive carboxypeptidase activity assigned it to carboxypeptidases A and B that are present in the posterior portion of the midgut (Houseman and Downe 1981; Terra et al. 1988). However, later analysis of transcriptome datasets revealed that hemipterans lack M14 enzymes, possibly associated with the loss of the digestive serine peptidases during the Hemiptera evolution (C. Ferreira et al. 2015). Thus, while readapting to protein-rich foods, hemipterans may also have recruited lysosomal carboxypeptidases, similar to what happened with peptidases (see items 4.2 and 4.3). *Dysdercus peruvianus* (Hemiptera) transcripts show the presence of two serine carboxypeptidases pertaining to family S10 (EC 3.4.16) with differential expression restricted to the midgut, with supporting data inferring secretion into the lumen and, therefore, appearing to be true digestive enzymes (C. Ferreira et al. 2015). Previous results had indicated the presence of similar enzymes in the midgut of *Triatoma brasiliensis* (Hemiptera), describing two serine carboxypeptidases coding genes, with the demonstration of luminal activity (Waniek et al. 2014).

Insects also feature transcripts for prolyl carboxypeptidases from family S28, found in all orders, except Dictyoptera. Several of the described sequences have a putative signal peptide. However, only one sequence was shown to be secreted, into the midgut lumen of *T. molitor* (i.e., Tm1935) (C. Ferreira et al. 2015). This data is in accordance with the previously observed prolyl carboxypeptidase activity purified and characterized from the anterior midgut of *T. molitor*, which was implicated in the hydrolysis of the insect's major dietary proteins, the gliadins, a component of gluten (Goptar et al. 2013). Furthermore, phylogenetic analysis of the encoding sequences for these enzymes suggest that they might originated after the divergence of the main insect orders (C. Ferreira et al. 2015).

Conclusion

Insect digestive peptidases are as diverse as the insects themselves and, in spite of the lack of available information; knowledge about these enzymes is remarkably growing and expanding. Noticeably and understandably, most of the work on insect digestive peptide hydrolases focus on the economic and medically important species, varying from the simple proteolytic activity record, to gene identification, sequencing, transcriptomic and phylogenetic analyses, as well as peptidase purification followed by full biochemical characterization. Undoubtedly, the fastest growing information relates to serine peptidases, especially trypsin, in a rate that is connected to the spread of this peptidase amongst insect species. The focus, however, seems to have shifted from the basic biological questions of the digestive machinery to the integrated possibilities of the development of new strategies of insect control, along with transmission preventing tools and the blockade of the development of several pathogensin their respective invertebrate hosts.

References

Abrudan, J., Ramalho-Ortigão, M., O'Neil, S., Stayback, G., Wadsworth, M., Bernard, M., Shoue, D., et al. (2013). "The Characterization of the Phlebotomus Papatasi Transcriptome." *Insect Molecular Biology*, *22* (2), 211–32.

Ahmad, Z., Saleemuddin, M. & Siddi, M. (1980). "Purification and Characterization of Three Alkaline Proteases from the Gut of the Larva of Army Worm, Spodoptera Litura." *Insect Biochemistry*, *10*, 667–73.

Almeida, L. C. P., de Souza, T. M., Tabosa, P. M. S., Soares, N. G., Rocha-Bezerra, L. C. B., Vasconcelos, I. M. & Carvalho, A. F. U. (2017). "Trypsin Inhibitor from Leucaena Leucocephala Seeds Delays and Disrupts the Development of Aedes Aegypti, a Multiple-Disease Vector." *Pest Management Science*, *73*(1), 181–87.

Almeida, L. C. P., Tabosa, P. M. S., Hissa, D. C., Vasconcelos, I. M. & Carvalho, A. F. U. (2018). "First Insights on Insecticidal Activity upon *Aedes Aegypti* and Partial Biochemical Characterization of a Novel Low Molecular Mass Chymotrypsin-Trypsin Inhibitor Purified from Lonchocarpus Sericeus Seeds." *Pest Management Science*, *74* (6), 1362–73.

Amiri, A., Bandani, A. R. & Alizadeh, H. (2016). "Molecular Identification Of Cysteine And Trypsin Protease, Effect Of Different Hosts On Protease Expression, And RNAi Mediated Silencing Of Cysteine Protease Gene In The Sunn Pest." *Archives of Insect Biochemistry and Physiology*, *91*(4), 189–209.

Ammunét, T., Klemola, N., Heisswolf, A. & Klemola, T. (2009). "Larval Parasitism of the Autumnal Moth Reduces Feeding Intensity on the Mountain Birch." *Plant-Animal Interactions*, *159*, 539–47.

Angleró-Rodríguez, Y. I., Talyuli, O. A. C., Blumberg, B. J., Kang, S., Demby, C., Shields, A., Carlson, J., Jupatanakul, N. & Dimopoulos, G. (2017). "An Aedes Aegypti-Associated Fungus Increases Susceptibility to Dengue Virus by Modulating Gut Trypsin Activity." *Microbiology and Infectios Disease*, *6*, 1–20.

Azambuja, P., Guimaraes, J. A. & Garcia, E. S. (1983). "Haemolytic Factor from the Crop of Rhodnius Prolixus: Evidence and Partial Characterization." *Journal of Insect Physiology*, *29*(11), 833–37.

Baker, J. E. & Woo, S. M. (1981). "Properties and Specificities of a Digestive Aminopeptidase from Larvae of Attagenus Megatoma (Coleoptera: Dermestidae)." *Comparative Biochemistry and Physiology Part B: Comparative Biochemistry*, *69*(2), 189–93.

Balczun, C., Pausch, J. K. & Schaub, G. A. (2012). "Blood Digestion in Triatomes - a Review." *Mitteilungen Der Deutschen Gesellschaft Für Allgemeine Und Angewandte Entomologie*, *18*, 331–34.

Balczun, C., Siemanowski, J., Pausch, J. K., Helling, S., Marcus, K., Stephan, C., Meyer, H. E., et al. (2012). "Intestinal Aspartate Proteases TiCatD and TiCatD2 of the Haematophagous Bug Triatoma Infestans (Reduviidae): Sequence Characterisation, Expression Pattern and Characterisation of Proteolytic Activity." *Insect Biochemistry and Molecular Biology*, *42*(4), 240–50.

Bansal, R. & Michel, A. (2018). "Expansion of Cytochrome P450 and Cathepsin Genes in the Generalist Herbivore Brown Marmorated Stink Bug." *BMC Genomics*, *19*(1), 1–14.

Barrett, A. J., Rawlings, N. D. & O'Brien, E. A. (2001). "The MEROPS Database as a Protease Information System." *Journal of Structural Biology*, *134*(2–3), 95–102.

Barrett, F. M. (1982). "Absorption of Fluid from the Anterior Midgut in Rhodnius." *Journal of Insect Physiology*, *28*(4), 335–41.

Bayes, A., Comellas-Bigler, M., de la Vega, M. R., Maskos, K., Bode, W., Aviles, F. X., Jongsma, M. A., Beekwilder, J. & Vendrell, J. (2005). "Structural Basis of the Resistance of an Insect Carboxypeptidase to Plant Protease Inhibitors." *Proceedings of the National Academy of Sciences*, *102*(46), 16602–7.

Beton, D., Guzzo, C. R., Ribeiro, A. F., Farah, C. S. & Walter, R. Terra. (2012). "The 3D Structure and Function of Digestive Cathepsin L-like Proteinases of Tenebrio Molitor Larval Midgut." *Insect Biochemistry and Molecular Biology*, *42*(9), 655–64.

Billingsley, P. F. & Downe, A. E. R. (1985). "Cellular Localisation of Aminopeptidase in the Midgut of Rhodnius Prolixus Stål (Hemiptera:Reduviidae) during Blood Digestion." *Cell and Tissue Research*, *241*(2), 421–28.

Borges-Veloso, A., Saboia-Vahia, L., Dias-Lopes, G., Domont, G. B., Britto, C., Cuervo, P. & de Jesus, J. B. (2015). "In-Depth Characterization of Trypsin-like Serine Peptidases in the Midgut of the Sugar Fed Culex Quinquefasciatus." *Parasites and Vectors*, *8*(1), 373.

Borges, E. C., Machado, E. M. M., Garcia, E. S. & Azambuja, P. (2006). "Trypanosoma Cruzi: Effects of Infection on Cathepsin D Activity in the Midgut of Rhodnius Prolixus." *Experimental Parasitology*, *112*(2), 130–33.

Botos, I., Meyer, E., Nguyen, M., Swanson, St. M., Koomen, J. M., Russell, D. H. & Meyer, E. F. (2000). "The Structure of an Insect Chymotrypsin." *Journal of Molecular Biology*, *298*(5), 895–901.

Bown, D. P., Wilkinson, H. S. & Gatehouse, J. A. (1997). "Differentially Regulated Inhibitor-Sensitive and Insensitive Protease Genes from the Phytophagous Insect Pest, Helicoverpa Armigera, Are Members of Complex Multigene Families." *Insect Biochemistry and Molecular Biology*, *27*(7), 625–38.

Božić, N., Ivanović, J., Nenadović, V., Bergström, J., Larsson, T. & Vujčić, Z. (2008). "Purification and Properties of Major Midgut Leucyl Aminopeptidase of Morimus Funereus (Coleoptera, Cerambycidae) Larvae." *Comparative Biochemistry and Physiology - B Biochemistry and Molecular Biology*, *149*(3), 454–62.

Bravo, A., Likitvivatanavong, S., Gill, S. S. & Soberón, M. (2011). "Bacillus Thuringiensis: A Story of a Successful Bioinsecticide." *Insect Biochemistry and Molecular Biology*, *41*(7), 423–31.

Broadway, R. M. (1997). "Dietary Regulation of Serine Proteinases That Are Resistant to Serine Proteinase Inhibitors." *Journal of Insect Physiology*, *43*(9), 855–74.

Buarque, D. S., Braz, G. R. C., Martins, R. M., Tanaka-Azevedo, A. M., Gomes, C. M., Oliveira, F. A. A., Schenkman, S. & Tanaka, A. S.

(2013). "Differential Expression Profiles in the Midgut of Triatoma Infestans Infected with Trypanosoma Cruzi." *PLoS ONE*, *8*(5), e61203.

Bulushova, N. V., Zhuzhikov, D. P., Lyutikova, L. I., Kirillova, N. E., Zalunin, I. A. & Chestukhina, G. G. (2011). "Toxin-Binding Proteins Isolated from Yellow Mealworm Tenebrio Molitor and Wax Moth Galleria Mellonella." *Biochemistry (Moscow)*, *76*(2), 202–8.

Chandra, G. S., Asokan, R., Manamohan, M., Ellango, R., Sharma, H. C., Akbar, S. M. D. & Kumar, N. K. K. (2017). "Double-Stranded RNA-Mediated Suppression of Trypsin-Like Serine Protease (t-SP) Triggers Over-Expression of Another t-SP Isoform in Helicoverpa Armigera." *Applied Biochemistry and Biotechnology*, *184*(2), 746–61.

Chen, H., Zhu, Y. C., Whitworth, R. J., Reese, J. C. & Chen, M. S. (2013). "Serine and Cysteine Protease-like Genes in the Genome of a Gall Midge and Their Interactions with Host Plant Genotypes." *Insect Biochemistry and Molecular Biology*, *43*(8), 701–11.

Chen, J., Likitvivatanavong, S., Aimanova, K. G. & Gill, S. S. (2013). "A 104kDa Aedes Aegypti Aminopeptidase N Is a Putative Receptor for the Cry11Aa Toxin from Bacillus Thuringiensis Subsp. Israelensis." *Insect Biochemistry and Molecular Biology*, *43*(12), 1201–08.

Cheung, A. C. & Gooding, R. H. (1970). "A Study of Trypsin in Calosoma Calidum (Fab.) and Carabus Taedatus Fab." *Comparative Biochemistry and Physiology*, *37*(3), 331–38.

Chikate, Y. R., Tamhane, V. A., Joshi, R. S., Gupta, V. S. & Giri, A. P. (2013). "Differential Protease Activity Augments Polyphagy in Helicoverpa Armigera." *Insect Molecular Biology*, *22*(3), 258–72.

Christeller, J. T. (2005). "Evolutionary Mechanisms Acting on Proteinase Inhibitor Variability." *FEBS Journal*, *272*(22), 5710–22.

Christeller, J. T., Laing, W. A., Shaw, B. D. & Burgess, E. P. J. (1990). "Characterization and Partial Purification of the Digestive Proteases of the Black Field Cricket, Teleogryllus Commodus (Walker): Elastase Is a Major Component." *Insect Biochemistry*, *20*(2), 157–64.

Christeller, J. T., Shaw, B. D., Gardiner, S. E. & Dymock, J. (1989). "Partial Purification and Characterization of the Major Midgut Proteases of

Grass Grub Larvae (Costelytra Zealandica, Coleoptera: Scarabaeidae)." *Insect Biochemistry*, *19*(3), 221–31.

Chu, C. C., Sun, W., Spencer, J. L., Pittendrigh, B. R. & Seufferheld, M. J. (2014). "Differential Effects of RNAi Treatments on Field Populations of the Western Corn Rootworm." *Pesticide Biochemistry and Physiology*, *110*(1), 1–6.

Cingel, A., Savić, J., Lazarević, J., Ćosić, T., Raspor, M., Smigocki, A. & Ninković, S. (2017). "Co-Expression of the Proteinase Inhibitors Oryzacystatin I and Oryzacystatin II in Transgenic Potato Alters Colorado Potato Beetle Larval Development." *Insect Science*, *24*(5), 768–80.

Cingel, A., Savić, J., Vinterhalter, B., Vinterhalter, D., Kostić, M., Jovanović, D. Š., Smigocki, A. & Ninković, S. (2015). "Growth and Development of Colorado Potato Beetle Larvae, Leptinotarsa Decemlineata, on Potato Plants Expressing the Oryzacystatin II Proteinase Inhibitor." *Transgenic Research*, *24*(4), 729–40.

Cohen, A. C. (1993). "Organization of Digestion and Preliminary Characterization of Salivary Trypsin-like Enzymes in a Predaceous Heteropteran, Zelus Renardii." *Journal of Insect Physiology*, *39*(10), 823–29.

Cohen, A. C. (1995). "Extra-Oral Digestion in Predaceous Terrestrial Arthropoda." *Annual Review of Entomology*, *40*(1), 85–103.

Cohen, A. C. & Wheeler, A. G. (1998). "Role of Saliva in the Highly Destructive Fourlined Plant Bug (Hemiptera: Miridae: Mirinae)." *Annals of the Entomological Society of America*, *91*(1), 94–100.

Colebatch, G., Cooper, P. & East, P. (2002). "CDNA Cloning of a Salivary Chymotrypsin-like Protease and the Identification of Six Additional CDNAs Encoding Putative Digestive Proteases from the Green Mirid, Creontiades Dilutus (Hemiptera: Miridae)." *Insect Biochemistry and Molecular Biology*, *32*(9), 1065–75.

Colebatch, G., East, P. & Cooper, P. (2001). "Preliminary Characterisation of Digestive Proteases of the Green Mirid, Creontiades Dilutus (Hemiptera: Miridae)." *Insect Biochemistry and Molecular Biology*, *31*(4–5), 415–23.

Colepicolo-Neto, P., Bechara, E. J. H., Ferreira, C. & Terra, W. R. (1986). "Evolutionary Considerations of the Spatial Organization of Digestion in the Luminescent Predaceous Larvae of Pyrearinus Termitilluminans (Coleoptera:Elateridae)." *Insect Biochemistry*, *16*(5), 811–17.

Costa, I. A., Samuels, R. I., Bifano, T. D., Terra, W. R. & Silva, C. P. (2011). "Purification and Partial Characterization of an Aminopeptidase from the Midgut Tissue of Dysdercus Peruvianus." *Comparative Biochemistry and Physiology - B Biochemistry and Molecular Biology*, *158*(3), 235–41.

Crava, C. M., Bel, Y., Jakubowska, A. K., Ferré, J. & Escriche, B. (2013). "Midgut Aminopeptidase N Isoforms from Ostrinia Nubilalis: Activity Characterization and Differential Binding to Cry1Ab and Cry1Fa Proteins from *Bacillus Thuringiensis*." *Insect Biochemistry and Molecular Biology*, *43*(10), 924–35.

Cristofoletti, P. T., De Sousa, F. A. M., Rahbé, Y. & Terra, W. R. (2006). "Characterization of a Membrane-Bound Aminopeptidase Purified from Acyrthosiphon Pisum Midgut Cells: A Major Binding Site for Toxic Mannose Lectins." *FEBS Journal*, *273*(24), 5574–88.

Cristofoletti, P. T., Ribeiro, A. F., Deraison, C., Rahbé, Y. & Terra, W. R. (2003). "Midgut Adaptation and Digestive Enzyme Distribution in a Phloem Feeding Insect, the Pea Aphid Acyrthosiphon Pisum." *Journal of Insect Physiology*, *49*(1), 11–24.

Cristofoletti, P. T., Ribeiro, A. F. & Terra, W. R. (2005). "The Cathepsin L-like Proteinases from the Midgut of Tenebrio Molitor Larvae: Sequence, Properties, Immunocytochemical Localization and Function." *Insect Biochemistry and Molecular Biology*, *35*(8), 883–901.

Cristofoletti, P. T. & Terra, W. R. (1999). "Specificity, Anchoring, and Subsites in the Active Center of a Microvillar Aminopeptidase Purified from Tenebrio Molitor (Coleoptera) Midgut Cells." *Insect Biochemistry and Molecular Biology*, *29*(9), 807–19.

Curzi, M. J., Zavala, J. A., Spencer, J. L. & Seufferheld, M. J. (2012). "Abnormally High Digestive Enzyme Activity and Gene Expression Explain the Contemporary Evolution of a Diabrotica Biotype Able to Feed on Soybeans." *Ecology and Evolution*, *2*(8), 2005–17.

Damasceno, T. F., Dias, R. O., de Oliveira, J. R., Salinas, R. K., Juliano, M. A., Ferreira, C. & Terra, W. R. (2017). "Active Subsite Properties, Subsite Residues and Targeting to Lysosomes or Midgut Lumen of Cathepsins L from the Beetle Tenebrio Molitor." *Insect Biochemistry and Molecular Biology*, *89*, 17–30.

Dana, A. N., Hong, Y. S., Kern, M. K., Hillenmeyer, M. E., Harker, B. W., Lobo, N. F., Hogan, J. R., Romans, P. & Collins, F. H. (2005). "Gene Expression Patterns Associated with Blood-Feeding in the Malaria Mosquito Anopheles Gambiae." *BMC Genomics*, *6*, 5.

Deraison, C., Darboux, I., Duportets, L., Gorojankina, T., Rahbé, Y. & Jouanin, L. (2004). "Cloning and Characterization of a Gut-Specific Cathepsin L from the Aphid Aphis Gossypii." *Insect Molecular Biology*, *13*(2), 165–77.

Dias-Lopes, G., Borges-Veloso, A., Saboia-Vahia, L., Domont, G. B., Britto, C., Cuervo, P. & De Jesus, J. B. (2015). "Expression of Active Trypsin-like Serine Peptidases in the Midgut of Sugar-Feeding Female Anopheles Aquasalis." *Parasites and Vectors*, *8*(1), 1–10.

Dias, R. O., Via, A., Brandão, M. M., Tramontano, A. & Silva-Filho, M. C. (2015). "Digestive Peptidase Evolution in Holometabolous Insects Led to a Divergent Group of Enzymes in Lepidoptera." *Insect Biochemistry and Molecular Biology*, *58*, 1–11.

Dunn, B. M., Jimenez, M., Parten, B. F., Valler, M. J., Rolph, C. E. & Kay, J. (1986). "A Systematic Series of Synthetic Chromophoric Substrates for Aspartic Proteinases." *The Biochemical Journal*, *237*(3), 899–906.

Edwards, M. J., Moskalyk, L. A., Donelly-Doman, M., Vlaskova, M., Noriega, F. G., Walker, V. K. & Jacobs-Lorena, M. (2000). "Characterization of a Carboxypeptidase a Gene from the Mosquito, *Aedes Aegypti*." *Insect Molecular Biology*, *9*(1), 33–38.

Espinoza-Fuentes, F. P. & Terra, W. R. (1987). "Physiological Adaptations for Digesting Bacteria. Water Fluxes and Distribution of Digestive Enzymes in Musca Domestica Larval Midgut." *Insect Biochemistry*, *17*(6), 809–17.

Felix, C. R., Betschart, B., Billingsley, P. F. & Freyvogel, T. A. (1991). "Post-Feeding Induction of Trypsin in the Midgut of Aedes Aegypti L.

(Diptera: Culicidae) Is Separable into Two Cellular Phases." *Insect Biochemistry*, *21*(2), 197–203.

Fernandes, K. M., de Magalhães-Júnior, M. J., Baracat-Pereira, M. C. & Martins, G. F. (2016). "Proteomic Analysis of Aedes Aegypti Midgut during Post-Embryonic Development and of the Female Mosquitoes Fed Different Diets." *Parasitology International*, *65*(6), 668–76.

Ferreira, A. H. P., Cristofoletti, P. T., Lorenzini, D. M., Guerra, L. O., Paiva, P. B., Briones, M. R. S., Terra, W. R. & Ferreira, C. (2007). "Identification of Midgut Microvillar Proteins from Tenebrio Molitor and Spodoptera Frugiperda by CDNA Library Screenings with Antibodies." *Journal of Insect Physiology*, *53*(11), 1112–24.

Ferreira, C., Rebola, K. G. O., Cardoso, C., Bragatto, I., Ribeiro, A. F. & Terra, W. R. (2015). "Insect Midgut Carboxypeptidases with Emphasis on S10 Hemipteran and M14 Lepidopteran Carboxypeptidases." *Insect Molecular Biology*, *24*(2), 222–29.

Ferreira, C., Ribeiro, A. F., Garcia, E. S. & Terra, W. R. (1988). "Digestive Enzymes Trapped between and Associated with the Double Plasma Membranes of Rhodnius Prolixus Posterior Midgut Cells." *Insect Biochemistry*, *18*(6), 521–30.

Ferreira, C. & Terra, W. R. (1989). "Spatial Organization of Digestion, Secretory Mechanisms and Digestive Enzyme Properties in Pheropsophus Aequinoctialis (Coleoptera: Carabidae)." *Insect Biochemistry*, *19*(4), 383–91.

Fialho, M. C. Q., Moreira, N. R., Zanuncio, J. C., Ribeiro, A. F., Terra, W. R. & Serrão, J. E. (2012). "Prey Digestion in the Midgut of the Predatory Bug Podisus Nigrispinus (Hemiptera: Pentatomidae)." *Journal of Insect Physiology*, *58*(6), 850–56.

Fialho, M. C. Q., Terra, W. R., Moreira, N. R., Zanuncio, J. C. & Serrão, J. E. (2013). "Ultrastructure and Immunolocalization of Digestive Enzymes in the Midgut of Podisus Nigrispinus (Heteroptera: Pentatomidae)." *Arthropod Structure and Development*, *42*(4), 277–85.

Fonseca, F. P. P., Soares-Costa, A., Ribeiro, A. F., Rosa, J. C., Terra, W. R. & Henrique-Silva, F. (2012). "Recombinant Expression, Localization and *in Vitro* Inhibition of Midgut Cysteine Peptidase (Sl-CathL) from

Sugarcane Weevil, Sphenophorus Levis." *Insect Biochemistry and Molecular Biology*, *42*(1), 58–69.

Franta, Z., Vogel, H., Lehmann, R., Rupp, O., Goesmann, A. & Vilcinskas, A. (2016). "Next Generation Sequencing Identifies Five Major Classes of Potentially Therapeutic Enzymes Secreted by Lucilia Sericata Medical Maggots." *BioMed Research International*, 2016, 8285428.

Galdeano, D. M., Breton, M. C., Lopes, J. R. S., Falk, B. W. & Machado, M. A. (2017). "Oral Delivery of Double-Stranded RNAs Induces Mortality in Nymphs and Adults of the Asian Citrus Psyllid, Diaphorina Citri." *PLoS ONE*, *12*(3), 1–15.

Garcia, E. S., Guimarães, J. A. & Prado, J. L. (1978). "Purification and Characterization of a Sulfhydryl-Dependent Protease from Rhodnius Prolixus Midgut." *Archives of Biochemistry and Biophysics*, *188*(2), 315–22.

Gatehouse, A. M. R., Norton, E., Davison, G. M., Babbé, S. M., Newell, C. A. & Gatehouse, J. A. (1999). "Digestive Proteolytic Activity in Larvae of Tomato Moth, Lacanobia Oleracea; Effects of Plant Protease Inhibitors *in Vitro* and *in Vivo*." *Journal of Insect Physiology*, *45*(6), 545–58.

Gatehouse, J. A. (2011). "Prospects for Using Proteinase Inhibitors to Protect Transgenic Plants Against Attack by Herbivorous Insects." *Current Protein & Peptide Science*, *12*(5), 409–16.

Gatehouse, L. N., Shannon, A. L., Burgess, E. P. J. & Christeller, J. T. (1997). "Characterization of Major Midgut Proteinase CDNAs from Helicoverpa Armigera Larvae and Changes in Gene Expression in Response to Four Proteinase Inhibitors in the Diet." *Insect Biochemistry and Molecular Biology*, *27*(11), 929–44.

Ge, Z. Y., Wan, P. J. & Han, Z. J. (2012). "Cloning and Characterization of Serpin-like Genes from the Striped Rice Stem Borer, Chilo Suppressalis." *Genome*, *55*, 359–66.

Giebel, W., Zwilling, R. & Pfleiderer, G. (1971). "The Evolution of Endopeptidases - XII . The Proteolytic Enzymes of the Honeybee (Apis Mellifica L .)." *Comparative Biochemistry and Physiology Part B: Comparative Biochemistry*, *38*(1), 197–210.

Giri, A., Harsulkar, A., Deshpande, V., Sainani, M., Gupta, V. & Ranjekar, P. (1998). "Chickpea Defensive Proteinase Inhibitors Can Be Inactivated by Podborer Gut Proteinases1." *Plant Physiology*, *116*(1), 393–401.

Gooding, R. H., Cheung, A. C. & Rolseth, B. M. (1973). "The Digestive Processes of Haematophagous Insects. III. Inhibition of Trypsin by Honey and the Possible Functions of the Oesophageal Diverticula of Mosquitoes (Diptera)." *The Canadian Entomologist*, *105*(3), 433–36.

Gooding, R. H. & Huang, C. T. (1969). "Trypsin and Chymotrypsin from the Beetle Pterostichus Melanarius." *Journal of Insect Physiology*, *15*(2), 325–39.

Gooding, R. H. & Rolseth, B. M. (1976). "Digestive Processes of Haematophagous Insects. XI Partial Purification and Some Properties of Six Proteolytic Enzymes from Tsetse Fly Glossina Morsitans Morsitans Westwood (Diptera: Glossinidae)." *Canadian Journal of Zoology*, *54*(11), 1950–59.

Goptar, I. A., Shagin, D. A., Shagina, I. A., Mudrik, E. S., Smirnova, Y. A., Zhuzhikov, D. P., Belozersky, M. A., et al. (2013). "A Digestive Prolyl Carboxypeptidase in Tenebrio Molitor Larvae." *Insect Biochemistry and Molecular Biology*, *43*(6), 501–9.

Graf, R., Boehlen, P. & Briegel, H. (1991). "Structural Diversity of Trypsin from Different Mosquito Species Feeding on Vertebrate Blood." *Experientia*, *47*(6), 603–9.

Graf, R. & Briegel, H. (1985). "Isolation of Trypsin Isozymes from the Mosquito Aedes Aegypti (L.)." *Insect Biochemistry*, *15*(5), 611–18.

Greenberg, B. & Paretsky, D. (1955). "Proteolytic Enzumes in the House Fly, Musca Domestica (L.)." *Annals*, *48*, 46–50.

Hartley, S. B. (1960). "Proteolytic Enzymes." *Annual Review of Biochemistry*, *29*, 45–72.

Harvey, J. A., Sano, T. & Tanaka, T. (2010). "Differential Host Growth Regulation by the Solitary Endoparasitoid, Meteorus Pulchricornis in Two Hosts of Greatly Differing Mass." *Journal of Insect Physiology*, *56*, 1178–83.

Henriques, B. S., Gomes, B., da Costa, S. G., Moraes, C. S., Mesquita, R. D., Dillon, V. M., Garcia, E. S., Azambuja, P., Dillon, R. J. & Genta, F. A. (2017). "Genome Wide Mapping of Peptidases in Rhodnius Prolixus: Identification of Protease Gene Duplications, Horizontally Transferred Proteases and Analysis of Peptidase A1 Structures, with Considerations on Their Role in the Evolution of Hematophagy in Triatomi." *Frontiers in Physiology*, *8*, 1–22.

Hoef, V. V., Breugelmans, B., Spit, J., Simonet, G., Zels, S. & Broeck, J. V. (2013). "Phylogenetic Distribution of Protease Inhibitors of the Kazal-Family within the Arthropoda." *Peptides*, *41*, 59–65.

Hooper, N. M. (2002). "Proteases : A Primer." *Essays in Biochemistry*, *38*, 1–8.

Hori, Koji. (1970). "Some Variations in the Activities of Salivary Amylase and Protease of Lygus Disponsi LINNAVUORI : Hemiptera : Miridae." *Applied Entomology and Zoology*, *5*(2), 51–61.

Hou, M. Z., Shen, G. M., Wei, D., Li, Y. L., Dou, W. & Wang, J. J. (2014). "Characterization of Bactrocera Dorsalis Serine Proteases and Evidence for Their Indirect Role in Insecticide Tolerance." *International Journal of Molecular Sciences*, *15*(2), 3272–86.

Houseman, J. G. (1978). "A Thiol-Activated Digestive Proteinase from Adults of Rhodnius Prolixus Stal (Hemiptera : Reduviidae)." *Canadian Journal of Zoology*, *1*, 1140–43.

Houseman, J. G. & Downe, A. E. R. (1980). "Endoproteinase Activity in the Posterior Midgut of Rhodnius Prolixus Stål (Hemiptera: Reduviidae)." *Insect Biochemistry*, *10*, 363–66.

Houseman, J. G. & Downe, A. E. R. (1981). "Exoproteinase Activity in the Posterior Midgut of Rhodnius Prolixus Stal (Hemiptera:Reduviidae)." *Insect Biochemistry*, *11*(5), 579–82.

Houseman, J. G. & Downe, A. E. R. (1982). "Characterization of an Acidic Proteinase from the Posterior Midgut of Rhodnius Prolixus Stål (Hemiptera: Reduviidae)." *Insect Biochemistry*, *12*(6), 651–55.

Houseman, J. G. & Downe, A. E. R. (1983). "Cathepsin D-like Activity in the Posterior Midgut of Hemipteran Insects." *Comparative*

Biochemistry and Physiology Part B: Comparative Biochemistry, *75*(3), 509–12.

Hu, X., Zhu, M., Wang, S., Zhu, L., Xue, R., Cao, G. & Gong, C. (2015). "Proteomics Analysis of Digestive Juice from Silkworm during Bombyx Mori Nucleopolyhedrovirus Infection." *Proteomics*, *15*(15), 2691–2700.

Huang, F., Cao, T. T., Min, S., Chen, Y. F. & Chen, X. X. (2008). "Parasitism-Induced Effects on Host Growth and Metabolic Efficiency in Plutella Xylostella Larvae Parasitized by Cotesia Vestalis or Diadegma Semiclausum." *Insect Science*, *15*(3), 237–43.

IUPAC-IUB Joint Commission on Biochemical Nomenclature (JCBN). 1984. "Nomenclature and Symbolism for Amino Acids and Peptides." *European Journal of Biochemistry*, *138*, 9–37.

Jany, K. D., Haug, H. & Ishay, J. (1978). "Trypsin-like Endopeptidases from the Midguts of the Larvae from the Hornets of Vespa Orientalis and Vespa Crabro." *Insect Biochemistry*, *8*, 221–30.

Jiang, J., Huang, Y., Shu, C., Soberón, M., Bravo, A., Liu, C., Song, F., Lai, J. & Zhang, J. (2017). "Holotrichia Oblita Midgut Proteins That Bind to Bacillus Thuringiensis Cry8-Like Toxin and Assembly of the H. Oblita Midgut Tissue Transcriptome." *Applied and Environmental Microbiology*, *83*(12), 1–11.

Johnson, K. S. & Rabosky, D. (2000). "Phylogenetic Distribution of Cysteine Proteinases in Beetles: Evidence for an Evolutionary Shift to an Alkaline Digestive Strategy in Cerambycidae." *Comparative Biochemistry and Physiology - B Biochemistry and Molecular Biology*, *126*(4), 609–19.

Johnston, K. A., Lee, M. J., Gatehouse, J. A. & Anstee, J. H. (1991). "The Partial Purification and Characterisation of Serine Protease Activity in Midgut of Larval Helicoverpa Armigera." *Insect Biochemistry*, *21*(4), 389–97.

Kaiser, L., Dupas, S., Branca, A., Herniou, E. A., Clarke, C. W., Dulac, C. C., Obonyo, J., et al. (2017). "The Cotesia Sesamiae Story: Insight into Host-Range Evolution in a Hymenoptera Parasitoid and Implication for Its Use in Biological Control Programs." *Genetica*, *145*(6), 455–68.

Kanost, M. R. & Clem, R. J. (2012). "Insect Proteases." In *Insect Molecular Biology and Biochemistry*, 2012, 346–64.

Klinkowstrom, A. M., Terra, W. R. & Ferreira, C. (1995). "Midgut Dipeptidases from Rhynchosciara Americana (Diptera) Larvae. Properties of Soluble and Membrane-Bound Forms." *Insect Biochemistry and Molecular Biology*, *25*(3), 303–10.

Knight, P. J., Knowles, B. H. & Ellar, D. J. (1995). "Molecular Cloning of an Insect Aminopeptidase N That Serves as a Receptor for Bacillus Thuringiensis CryIA(c) Toxin." *The Jounal of Biological Chemistry*, *270*(30), 17765–70.

Kumar, A., Sharma, A., Sharma, R. & Gakhar, S. K. (2014). "Identification, Characterization and Analysis of Expression of Gene Encoding Carboxypeptidase A in Anopheles Culicifacies A (Diptera: Culicidae)." *Acta Tropica*, *139*, 123–30.

Kuwar, S. S., Pauchet, Y., Vogel, H. & Heckel, D. G. (2015). "Adaptive Regulation of Digestive Serine Proteases in the Larval Midgut of Helicoverpa Armigera in Response to a Plant Protease Inhibitor." *Insect Biochemistry and Molecular Biology*, *59*, 18–29.

Lavazec, C., Boudin, C., Lacroix, R., Bonnet, S., Diop, A., Thiberge, S., Boisson, B., Tahar, R. & Bourgouin, C. (2007). "Carboxypeptidases B of Anopheles Gambiae as Targets for a Plasmodium, Falciparum Transmission-Blocking Vaccine." *Infection and Immunity*, *75*(4), 1635–42.

Lee, S. B., Aimanova, K. G. & Gill, S. S. (2014). "Alkaline Phosphatases and Aminopeptidases Are Altered in a Cry11Aa Resistant Strain of Aedes Aegypti." *Insect Biochemistry and Molecular Biology*, *54*, 112–21.

Lemos, F. J. A. & Terra, W. R. (1992). "Soluble and Membrane-Bound Forms of Trypsin-like Enzymes in Musca Domestica Larval Midguts." *Insect Biochemistry and Molecular Biology*, *22*(7), 613–19.

Lemos, F. & Terra, W. R. (1991). "Properties and Intracellular Distribution of a Cathepsin D-like Proteinase Active at the Acid Region of Musca Domestica Midgut." *Insect Biochemistry*, *21*(5), 457–65.

Levine, E. & Oloumi-Sadeghi, H. (1996). "Western Corn Rootworm (Coleoptera: Chrysomelidae) Larval Injury to Corn Grown for Seed Production Following Soybeans Grown for Seed Production." *Journal of Economic Entomology*, *89*(4), 1010–16.

Levine, E., Spencer, J. L., Isard, S. A., Onstad, D. W. & Gray, M. E. (2002). "Adaptation of the Western Corn Rootworm to Crop Rotation: Evolution of a New Strain in Response to a Management Practice." *American Entomologist*, *48*(2), 94–107.

Levinsky, H., Birk, Y. & Applebaum, S. W. (1977). "Isolation and Characteriztion of a New Trypsin-like Enzyme from Tenebrio Molitor L. Larvae." *International Journal of Peptide and Protein Research*, *10*, 252–64.

Li, H., Leavengood, J. M., Chapman, E. G., Burkhardt, D., Song, F., Jiang, P., Liu, J., Zhou, X. & Cai, W. (2017). "Mitochondrial Phylogenomics of Hemiptera Reveals Adaptive Innovations Driving the Diversification of True Bugs." *Proceedings of the Royal Society B: Biological Sciences*, *284*(1862), 20171223.

Li, Y. L., Hou, M. Z., Shen, G. M., Lu, X. P., Wang, Z., Jia, F. X., Wang, J. J. & Dou, W. (2017). "Functional Analysis of Five Trypsin-like Protease Genes in the Oriental Fruit Fly, Bactrocera Dorsalis (Diptera: Tephritidae)." *Pesticide Biochemistry and Physiology*, *136*, 52–57.

Lima, T. A., Pontual, E. V., Dornelles, L. P., Amorim, P. K., Sá, R. A., Coelho, L. C. B. B., Napoleão, T. H. & Paiva, P. M. G. (2014). "Digestive Enzymes from Workers and Soldiers of Termite Nasutitermes Corniger." *Comparative Biochemistry and Physiology Part B: Biochemistry and Molecular Biology*, *176*(1), 1–8.

Liu, S., Lomate, P. R. & Bonning, B. C. (2018). "Tissue-Specific Transcription of Proteases and Nucleases across the Accessory Salivary Gland, Principal Salivary Gland and Gut of Nezara Viridula." *Insect Biochemistry and Molecular Biology*, *21*(103), 36–45.

Lopes, A. R., Juliano, M. A., Juliano, L. & Terra, W. R. (2004). "Coevolution of Insect Trypsins and Inhibitors." *Archives of Insect Biochemistry and Physiology*, *55*(3), 140–52.

Lu, S. J., Pennington, J. E., Stonehouse, A. R., Mobula, M. M. & Wells, M. A. (2006). "Reevaluation of the Role of Early Trypsin Activity in the Transcriptional Activation of the Late Trypsin Gene in the Mosquito Aedes Aegypti." *Insect Biochemistry and Molecular Biology*, *36*(4), 336–43.

Macedo, M. L. R. & Freire, M. G. M. (2011). "Insect Digestive Enzymes as a Target for Pest Control." *Invertebrate Survival Journal*, *8*, 190–98.

Mahajan, N. S., Mishra, M., Tamhane, Va. A., Gupta, V. S. & Giri, A. P. (2013). "Plasticity of Protease Gene Expression in Helicoverpa Armigera upon Exposure to Multi-Domain Capsicum Annuum Protease Inhibitor." *Biochimica et Biophysica Acta - General Subjects*, *1830*(6), 3414–20.

Mahdavi, A., Ghadamyari, M., Sajedi, R. H., Sharifi, M. & Kouchaki, B. (2013). "Identification and Partial Characterization of Midgut Proteases in the Lesser Mulberry Pyralid, *Glyphodes Pyloalis*." *Journal of Insect Science*, *13*(81), 1–11.

Malaikozhundan, B. & Vinodhini, J. (2018). "Biological Control of the Pulse Beetle, Callosobruchus Maculatus in Stored Grains Using the Entomopathogenic Bacteria, Bacillus Thuringiensis." *Microbial Pathogenesis*, *114*, 139–46.

Martin, M. M. (1983). "Cellulose Digestion in Insects." *Comparative Biochemistry and Physiology Part A: Physiology*, *75*(3), 313–24.

Martinez, M., Santamaria, M. E., Diaz-Mendoza, M., Arnaiz, A., Carrillo, L. & Ortego, F. (2016). "Phytocystatins: Defense Proteins against Phytophagous Insects and Acari." *International Journal of Molecular Sciences*, *17*(10), 1–16.

Martynov, A. G., Elpidina, E. N., Perkin, L. & Oppert, B. (2015). "Functional Analysis of C1 Family Cysteine Peptidases in the Larval Gut of Tenebrio Molitor and Tribolium Castaneum." *BMC Genomics*, *16* (1), 75.

Mazumdar-Leighton, S. & Broadway, R. M. (2001a). "Identification of Six Chymotrypsin CDNAs from Larval Midguts of Helicoverpa Zea and Agrotis Ipsilon Feeding on the Soybean (Kunitz) Trypsin Inhibitor." *Insect Biochemistry and Molecular Biology*, *31*, 633–44.

Mazumdar-Leighton, S. & Broadway, R. M. (2001b). "Transcriptional Induction of Diverse Midgut Trypsins in Larval Agrotis Ipsilon and Helicoverpa Zea Feeding on the Soybean Trypsin Inhibitor." *Insect Biochemistry and Molecular Biology*, *31*, 645–57.

Medel, V., Palma, R., Mercado, D., Rebolledo, R., Quiroz, A. & Mutis, A. (2015). "The Effect of Protease Inhibitors on Digestive Proteolytic Activity in the Raspberry Weevil, Aegorhinus Superciliosus (Guérin) (Coleoptera: Curculionidae)." *Neotropical Entomology*, *44*(1), 77–83.

Meekins, D. A., Kanost, M. R. & Michel, K. (2017). "Serpins in Arthropod Biology." *Seminars in Cell & Developmental Biology*, *62*, 105–19.

Mehrabadi, M., Bandani, A. R. & Dastranj, M. (2014). "Salivary Digestive Enzymes of the Wheat Bug, Eurygaster Integriceps (Insecta: Hemiptera: Scutelleridae)." *Comptes Rendus – Biologies*, *337*(6), 373–82.

Mesquita, R. D., Vionette-Amaral, R. J., Lowenberger, C., Rivera-Pomar, R., Monteiro, F. A., Minx, P., Spieth, J., et al. (2015). "Genome of Rhodnius Prolixus, an Insect Vector of Chagas Disease, Reveals Unique Adaptations to Hematophagy and Parasite Infection." *Proceedings of the National Academy of Sciences*, *112*(48), 14936–41.

Miller, J. W., Kramer, K. J. & Law, J. H. (1974). "Isolation and Partial Characterization of the Larval Midgut Trypsin from the Tobacco Hornworm, Manduca Sexta Johannson (Lepidoptera: Sphingidale)." *Comparative Biochemistry and Physiology Part B: Comparative Biochemistry*, *48* (1), 117–20.

Milne, R. & Kaplan, H. (1993). "Purification and Characterization of a Trypsin-like Digestive Enzyme from Spruce Budworm (Choristoneura Fumiferana) Responsible for the Activation of CS-Endotoxin from Bacillus Thuringiensis." *Insect Biochemistry and Molecular Biology*, *23*(6), 663–73.

Misof, B., Liu, S., Meusemann, K., Peters, R. S., Donath, A., Mayer, C., Frandsen, P. B., et al. (2014). "Phylogenomics Resolves the Timing and Pattern of Insect Evolution." *Science*, *346*(6210), 763–67.

Molina, D., Patiño, L., Quintero, M., Cortes, J. & Bastos, S. (2014). "Effects of the Aspartic Protease Inhibitor from Lupinus Bogotensis Seeds on the Growth and Development of Hypothenemus Hampei: An Inhibitor

Showing High Homology with Storage Proteins." *Phytochemistry*, *98*, 69–77.

Monteiro, F. A., Weirauch, C., Felix, M., Lazoski, C. & Abad-Franch, F. (2018). "Evolution, Systematics, and Biogeography of the Triatominae, Vectors of Chagas Disease." *Advances in Parasitology*, *99*, 265–344.

Moraes, C. S., Aguiar-Martins, K., Costa, S. G., Bates, P. A., Dillon, R. J. & Genta, F. A. (2018). "Second Blood Meal by Female Lutzomyia Longipalpis: Enhancement by Oviposition and Its Effects on Digestion, Longevity, and Leishmania Infection." *BioMed Research International*, 2018, 1–10.

Mucha, A., Drag, M., Dalton, J. P. & Kafarski, P. (2010). "Metallo-Aminopeptidase Inhibitors." *Biochimie*, *92*(11), 1509–29.

Munster, M. V., Le Gleuher, M., Pauchet, Y., Augustin, S., Courtin, C., Amichot, M., French-Constant, R. H. & Pauron, D. (2011). "Molecular Characterization of Three Genes Encoding Aminopeptidases N in the Poplar Leaf Beetle Chrysomela Tremulae." *Insect Molecular Biology*, *20*(2), 267–78.

Nakanishi, K., Yaoi, K., Nagino, Y., Hara, H., Kitami, M., Atsumi, S., Miura, N. & Sato, R. (2002). "Aminopeptidase N Isoforms from the Midgut of Bombyx Mori and Plutella Xylostella - Their Classification and the Factors That Determine Their Binding Specificity to Bacillus Thuringiensis Cry1A Toxin." *FEBS Letters*, *519*(1–3), 215–20.

Navaei-Bonab, R., Kazzazi, M., Saber, M. & Vatanparast, M. (2018). "Differential Inhibition of Helicoverpa Armigera (Lep.: Noctuidae) Gut Digestive Trypsin by Extracted and Purified Inhibitor of Datura Metel (Solanales: Solanaceae)." *Journal of Economic Entomology*, *111*(1), 178–86.

Noriega, F. G., Wang, X. Y., Pennington, J. E., Barillas-Mury, C. V. & Wells, M. A. (1996). "Early Trypsin, a Female-Specific Midgut Protease in Aedes Aegypti: Isolation, Amino-Terminal Sequence Determination, and Cloning and Sequencing of the Gene." *Insect Biochemistry and Molecular Biology*, *26*(2), 119–26.

Noriega, F. G. & Wells, M. A. (1999). "A Molecular View of Trypsin Synthesis in the Midgut of Aedes Aegypti." *Journal of Insect Physiology*, *45*(7), 613–20.

Okuda, K., Caroci, A., Ribolla, P., Marinotti, O., De Bianchi, A. G. & Bijovsky, A. T. (2005). "Morphological and Enzymatic Analysis of the Midgut of Anopheles Darlingi during Blood Digestion." *Journal of Insect Physiology*, *51*(7), 769–76.

Oliveira, C. F. R. de., Marangoni, S. & Macedo, M. L. R. (2014). "The Trypsin Inhibitor from Entada Acaciifolia Seeds Affects Negatively the Development of Mediterranean Flour Moth, Anagasta Kuehniella." *Pesticide Biochemistry and Physiology*, *108*(1), 74–79.

Oliveira, C. F. R. de., Souza, T. P., Parra, J. R. P., Marangoni, S., Silva-Filho, M. C. & Macedo, M. L. R. (2013). "Insensitive Trypsins Are Differentially Transcribed during Spodoptera Frugiperda Adaptation against Plant Protease Inhibitors." *Comparative Biochemistry and Physiology - B Biochemistry and Molecular Biology*, *165*(1), 19–25.

Oppert, B., Kramer, K. J., Beeman, R. W., Johnson, D. & McGaughey, W. H. (1997). "Proteinase-Mediated Insect Resistance to Bacillus Thuringiensis Toxins." *The Journal of Biological Chemistry*, *272*(38), 23473–76.

Orona-Tamayo, D., Wielsch, N., Blanco-Labra, A., Svatos, A., Farías-Rodríguez, R. & Heil, M. (2013). "Exclusive Rewards in Mutualisms: Ant Proteases and Plant Protease Inhibitors Create a Lock-Key System to Protect Acacia Food Bodies from Exploitation." *Molecular Ecology*, *22*(15), 4087–4100.

Padilha, M. H. P., Pimentel, A. C., Ribeiro, A. F. & Terra, W. R. (2009). "Sequence and Function of Lysosomal and Digestive Cathepsin D-like Proteinases of Musca Domestica Midgut." *Insect Biochemistry and Molecular Biology*, *39*(11), 782–91.

Patil, C. D., Borase, H. P., Suryawanshi, R. K. & Patil, S. V. (2016). "Trypsin Inactivation by Latex Fabricated Gold Nanoparticles: A New Strategy towards Insect Control." *Enzyme and Microbial Technology*, *92*, 18–25.

Pauchet, Y., Wilkinson, P., van Munster, M., Augustin, S., Pauron, D. & French-Constant, R. H. (2009). "Pyrosequencing of the Midgut Transcriptome of the Poplar Leaf Beetle Chrysomela Tremulae Reveals New Gene Families in Coleoptera." *Insect Biochemistry and Molecular Biology*, *39*(5–6), 403–13.

Perkin, L. C., Elpidina, E. N. & Oppert, B. (2017). "RNA Interference and Dietary Inhibitors Induce a Similar Compensation Response in Tribolium Castaneum Larvae." *Insect Molecular Biology*, *26*(1), 35–45.

Peterson, A. M., Barillas-Mury, C. V. & Wells, M. A. (1994). "Sequence of Three CDNAs Encoding an Alkaline Midgut Trypsin from Manduca Sexta." *Insect Biochemistry and Molecular Biology*, *24*(5), 463–71.

Peterson, A. M., Fernando, G. J. P. & Wells, M. A. (1995). "Purification, Characterization and CDNA Sequence of an Alkaline Chymotrypsin from the Midgut of Manduca Sexta." *Insect Biochemistry and Molecular Biology*, *25*(7), 765–74.

Pigott, C. R. & Ellar, D. J. (2007). "Role of Receptors in Bacillus Thuringiensis Crystal Toxin Activity." *Microbiology and Molecular Biology Reviews*, *71*(2), 255–81.

Pimenta, D. C., Oliveira, A., Juliano, M. A. & Juliano, L. (2001). "Substrate Specificity of Human Cathepsin D Using Internally Quenched Fluorescent Peptides Derived from Reactive Site Loop of Kallistatin." *Biochimica et Biophysica Acta - Protein Structure and Molecular Enzymology*, *1544*(1–2), 113–22.

Pimenta de Oliveira, S., de Oliveira, C. D., Sant'Anna, M. R. V., Dutra, H. L. C., Caragata, E. P. & Moreira, L. A. (2017). "Wolbachia Infection in Aedes Aegypti Mosquitoes Alters Blood Meal Excretion and Delays Oviposition without Affecting Trypsin Activity." *Insect Biochemistry and Molecular Biology*, *87*, 65–74.

Pimentel, A. C., Fuzita, F. J., Palmisano, G., Ferreira, C. & Terra, W. R. (2017). "Role of Cathepsins D in the Midgut of Dysdercus Peruvianus." *Comparative Biochemistry and Physiology Part B: Biochemistry and Molecular Biology*, *204*, 45–52.

Ponnuvel, K. M., Nithya, K., Sirigineedi, S., Awasthi, A. K. & Yamakawa, M. (2012). "*In Vitro* Antiviral Activity of an Alkaline Trypsin from the

Digestive Juice of Bombyx Mori Larvae against Nucleopolyhedrovirus." *Archives of Insect Biochemistry and Physiology*, *81*(2), 90–104.

Pruzinova, K., Votypka, J. & Volf, P. (2013). "The Effect of Avian Blood on Leishmania Development in Phlebotomus Duboscqi." *Parasites and Vectors*, *6* (1), 254.

Rajagopal, R., Agrawal, N., Selvapandiyan, A., Sivakumar, S., Ahmad, S. & Bhatnagar, R. K. (2003). "Recombinantly Expressed Isoenzymic Aminopeptidases from Helicoverpa Armigera (American Cotton Bollworm) Midgut Display Differential Interaction with Closely Related Bacillus Thuringiensis Insecticidal Proteins." *Biochemical Journal*, *978* (370), 971–78.

Ranjbar, M., Zibaee, A. & Sendi, J. J. (2014). "A Trypsin-like Proteinase in the Midgut of Ectomyelois Ceratoniae Zeller (Lepidoptera: Pyralidae): Purification, Characterization, and Host Plant Inhibitors." *Archives of Insect Biochemistry and Physiology*, *85*(1), 1–12.

Rasoolizadeh, A., Goulet, M. C., Guay, J. F., Cloutier, C. & Michaud, D. (2018). "Population-Associated Heterogeneity of the Digestive Cys Protease Complement in Colorado Potato Beetle, Leptinotarsa Decemlineata." *Journal of Insect Physiology*, *106*, 125–33.

Rasoolizadeh, A., Goulet, M. C., Sainsbury, F., Cloutier, C. & Michaud, D. (2016). "Single Substitutions to Closely Related Amino Acids Contribute to the Functional Diversification of an Insect-Inducible, Positively Selected Plant Cystatin." *FEBS Journal*, *283*(7), 1323–35.

Rawlings, N. D. & Barrett, A. J. (1993). "Evolutionary Families of Peptidases." *Biochemical Journal*, *290*, 205–18.

Rawlings, N. D., Barrett, A. J. & Bateman, A. (1999). "MEROPS: The Peptidase Database." *Nucleic Acid Research*, *27* (1), 325–31.

Rawlings, N. D. & Salvesen, G. (2013). *Handbook of Proteolytic Enzymes. Elsevier*. 3rd edition.

Raz, A., Djadid, N. D. & Zakeri, S. (2013). "Molecular Characterization of the Carboxypeptidase B1 of Anopheles Stephensi and Its Evaluation as a Target for Transmission-Blocking Vaccines." *Infection and Immunity*, *81*(6), 2206–16.

Reid, W. R., Zhang, L. & Liu, N. (2015). "Temporal Gene Expression Profiles of Pre Blood-Fed Adult Females Immediately Following Eclosion in the Southern House Mosquito Culex Quinquefasciatus." *International Journal of Biological Sciences*, *11*(11), 1306–13.

Rezende, T. M. T., Romão, T. P., Batista, M., Berry, C., Adang, M. J. & Silva-Filha, M. H. N. L. (2017). "Identification of Cry48Aa/Cry49Aa Toxin Ligands in the Midgut of Culex Quinquefasciatus Larvae." *Insect Biochemistry and Molecular Biology*, *88*, 63–70.

Ribeiro, J. M. C., Genta, F. A., Sorgine, M. H. F., Logullo, R., Mesquita, R. D., Paiva-Silva, G. O., Majerowicz, D., et al. (2014). "An Insight into the Transcriptome of the Digestive Tract of the Bloodsucking Bug, Rhodnius Prolixus." *PLoS Neglected Tropical Diseases*, *8*(1), e2594.

Rossi, G. D., Salvador, G. & Cônsoli, F. L. (2014). "The Parasitoid, Cotesia Flavipes (Cameron) (Hymenoptera: Braconidae), Influences Food Consumption and Utilization by Larval Diatraea Saccharalis (F.) (Lepidoptera: Crambidae)." *Archives of Insect Biochemistry and Physiology*, *87*(2), 85–94.

Ryan, C. A. (1990). "Protease Inhibitors in Plants: Genes for Improving Defenses Against Insects and Pathogens." *Annual Review of Phytopathology*, *28*(1), 425–49.

Saadati, F. & Bandani, A. R. (2011). "Effects of Serine Protease Inhibitors on Growth and Development and Digestive Serine Proteinases of the Sunn Pest, Eurygaster Integriceps." *Journal of Insect Science*, *11*(72), 1–12.

Saboia-Vahia, L., Borges-Veloso, A., Mesquita-Rodrigues, C., Cuervo, P., Dias-Lopes, G., Britto, C., De Barros Silva, A. P. & De Jesus, J. B. (2013). "Trypsin-like Serine Peptidase Profiles in the Egg, Larval, and Pupal Stages of Aedes Albopictus." *Parasites and Vectors*, *6*(1), 1–11.

Saboia-Vahia, L., Cuervo, P., Borges-Veloso, A., De Souza, N. P., Britto, C., Dias-Lopes, G. & De Jesus, J. B. (2014). "The Midgut of Aedes Albopictus Females Expresses Active Trypsin-like Serine Peptidases." *Parasites and Vectors*, *7*(1), 1–8.

Sakal, E., Applebaum, S. W. & Birk, Y. (1989). "Purification and Characterization of Trypsins from the Digestive Tract of Locusta

Migratoria." *International Journal of Peptide and Protein Research*, *34*(6), 498–505.

Salvador, G. & Cônsoli, F. L. (2008). "Changes in the Hemolymph and Fat Body Metabolites of Diatraea Saccharalis (Fabricius) (Lepidoptera: Crambidae) Parasitized by Cotesia Flavipes (Cameron) (Hymenoptera: Braconidae)." *Biological Control*, *45*(1), 103–10.

Santos, C. D., Ferreira, C. & Terra, W. R. (1983). "Consumption of Food and Spatial Organization of Digestion in the Cassava Hornworm, Erinnyis Ello." *Journal of Insect Physiology*, *29*(9), 707–14.

Santos, V. C., Vale, V. F., Silva, S. M., Nascimento, A. A. S., Saab, N. A. A., Soares, R. P. P., Michalick, M. S. M., et al. (2014). "Host Modulation by a Parasite: How Leishmania Infantum Modifies the Intestinal Environment of Lutzomyia Longipalpis to Favor Its Development." *PLoS ONE*, *9*(11), 1–9.

Schaub, G. A., Vogel, P. & Balczun, C. (2016). "Parasite-Vector Interactions." In *Molecular Parasitology*, edited by J. Walochnik and M. Duchêne, 431–89. Springer-Verlag Wien.

Schechter, I. & Berger, A. (1967). "On the Size of the Active Site in Proteases. 1. Papain." *Biochemical and Biophysical Research Communications*, *27*(2), 157–62.

Schneider, M., Rudin, W. & Hecker, H. (1986). "Absorption and Transport of Radioactive Tracers in the Midgut of the Malaria Mosquito, Anopheles Stephensi." *Journal of Ultrastructure Research and Molecular Structure Research*, *97*(1–3), 50–63.

Schumaker, T. T. S., Cristofoletti, P. T. & Terra, W. R. (1993). "Properties and Compartmentalization of Digestive Carbohydrases and Proteases in Scaptotrigona Bipunctata (Apidae, Meliponinae) Larvae." *Apidologie*, *24*(1), 3–17.

Sellami, S. & Jamoussi, K. (2016). "Investigation of Larvae Digestive β-Glucosidase and Proteases of the Tomato Pest Tuta Absoluta for Inhibiting the Insect Development." *Bulletin of Entomological Research*, *106*(3), 406–14.

Sethi, A., Xue, Q. G., La Peyre, J. F., Delatte, J. & Husseneder, C. (2011). "Dual Origin of Gut Proteases in Formosan Subterranean Termites

(Coptotermes Formosanus Shiraki) (Isoptera: Rhinotermitidae)." *Comparative Biochemistry and Physiology - A Molecular and Integrative Physiology*, *159*(3), 261–67.

Sharifi, M., Chitgar, M. G., Ghadamyari, M. & Ajamhasani, M. (2012). "Identification and Characterization of Midgut Digestive Proteases From the Rosaceous Branch Borer, Osphranteria Coerulescens Redtenbacher (Coleoptera : Cerambycidae)." *Romanian Journal of Biochemistry*, *49*(1), 33–47.

Sharifloo, A., Zibaee, A., Sendi, J. J. & Jahroumi, K. T. (2018). "Biochemical Characterization a Digestive Trypsin in the Midgut of Large Cabbage White Butterfly, Pieris Brassicae L. (Lepidoptera: Pieridae)." *Bulletin of Entomological Research*, *108*(4), 501–9.

Shen, G. M., Dou, W., Huang, Y., Jiang, X. Z., Smagghe, G. & Wang, J. J. (2013). "In Silico Cloning and Annotation of Genes Involved in the Digestion, Detoxification and RNA Interference Mechanism in the Midgut of Bactrocera Dorsalis [Hendel (Diptera: Tephritidae)]." *Insect Molecular Biology*, *22*(4), 354–65.

Shi, M., Zhu, N., Yi, Y. & Chen, X. X. (2013). "Four Serine Protease Cdnas from the Midgut of Plutella Xylostella and Their Proteinase Activity Are Influenced by the Endoparasitoid, Cotesia Vestalis." *Archives of Insect Biochemistry and Physiology*, *83*(2), 101–14.

Shu, C., Tan, S., Yin, J., Soberón, M., Bravo, A., Liu, C., Geng, L., Song, F., Li, K. & Zhang, J. (2015). "Assembling of Holotrichia Parallela (Dark Black Chafer) Midgut Tissue Transcriptome and Identification of Midgut Proteins That Bind to Cry8Ea Toxin from Bacillus Thuringiensis." *Applied Microbiology and Biotechnology*, *99*(17), 7209–18.

Silva, C. P. & Terra, W. R. (1994). "Digestive and Absorptive Sites along the Midgut of the Cotton Seed Sucker Bug Dysdercus Peruvianus (Hemiptera: Pyrrhocoridae)." *Insect Biochemistry and Molecular Biology*, *24*(5), 493–505.

Silva, W., Cardoso, C., Ribeiro, A. F., Terra, W. R. & Ferreira, C. (2013). "Midgut Proteins Released by Microapocrine Secretion in Spodoptera Frugiperda." *Journal of Insect Physiology*, *59*(1), 70–80.

Simpson, R. M., Poulton, J. & Markwick, N. P. (2008). "Expression Levels of Aminopeptidase-N Genes in the Lightbrown Apple Moth, Epiphyas Postvittana." *Insect Science*, *15*(6), 505–12.

Smilanich, A. M., Mason, P. A., Sprung, L., Chase, T. R. & Singer, M. S. (2011). "Complex Effects of Parasitoids on Pharmacophagy and Diet Choice of a Polyphagous Caterpillar." *Oecologia*, *165*(4), 995–1005.

Soares, T. S., Gonzalez, B. L. R., Torquato, R. J. S., Lemos, F. J. A., Costa-da-Silva, A. L., Guimarães, M. L. C. & Tanaka, A. S. (2018). "Functional Characterization of a Serine Protease Inhibitor Modulated in the Infection of the Aedes Aegypti with Dengue Virus." *Biochimie*, *144*, 160–68.

Soares, T. S., Torquato, R. J. S., Gonzalez, Ya., Lemos, F. J. A. & Tanaka, A. S. (2015). "Production of Serine Protease Inhibitors by Mutagenesis and Their Effects on the Mortality of Aedes Aegypti L. Larvae." *Parasites and Vectors*, *8*(1), 1–8.

Soares, T. S., Torquato, R. J. S., Lemos, F. J. A. & Tanaka, A. S. (2013). "Selective Inhibitors of Digestive Enzymes from Aedes Aegypti Larvae Identified by Phage Display." *Insect Biochemistry and Molecular Biology*, *43*(1), 9–16.

Soares, T. S., Watanabe, R. M. O., Lemos, F. J. A. & Tanaka, A. S. (2011). "Molecular Characterization of Genes Encoding Trypsin-like Enzymes from Aedes Aegypti Larvae and Identification of Digestive Enzymes." *Gene*, *489*(2), 70–75.

Souza, T. P., Dias, R. O., Castelhano, E. C., Brandão, M. M., Moura, D. S. & Silva-Filho, M. C. (2016). "Comparative Analysis of Expression Profiling of the Trypsin and Chymotrypsin Genes from Lepidoptera Species with Different Levels of Sensitivity to Soybean Peptidase Inhibitors." *Comparative Biochemistry and Physiology Part B: Biochemistry and Molecular Biology*, 196–197, 67–73.

Spit, J., Zels, S., Dillen, S., Holtof, M., Wynant, N. & Broeck, J. V. (2014). "Effects of Different Dietary Conditions on the Expression of Trypsin- and Chymotrypsin-like Protease Genes in the Digestivesystem of the Migratory Locust, Locusta Migratoria." *Insect Biochemistry and Molecular Biology*, *48*(1), 100–109.

Sriwichai, P., Rongsiryam, Y., Jariyapan, N., Sattabongkot, J., Apiwathnasorn, C., Nacapunchai, D. & Paskewitz, S. (2012). "Cloning of a Trypsin-like Serine Protease and Expression Patterns during Plasmodium Falciparum Invasion in the Mosquito, Anopheles Dirus (Peyton and Harrison)." *Archives of Insect Biochemistry and Physiology*, *80*(3), 151–65.

Srp, J., Nussbaumerová, M., Horn, M. & Mareš, M. (2016). "Digestive Proteolysis in the Colorado Potato Beetle, Leptinotarsa Decemlineata: Activity-Based Profiling and Imaging of a Multipeptidase Network." *Insect Biochemistry and Molecular Biology*, *78*, 1–11.

Stevens, J. A., Dunse, K. M., Guarino, R. F., Barbeta, B. L., Evans, S. C., West, J. A. & Anderson, M. A. (2013). "The Impact of Ingested Potato Type II Inhibitors on the Production of the Major Serine Proteases in the Gut of Helicoverpa Armigera." *Insect Biochemistry and Molecular Biology*, *43*(2), 197–208.

Takanona, T. & Hori, K. (1974). "Digestive Enzymes in the Salivary Gland and Midgut of the Bug Stenotus Binotatus." *Comparative Biochemistry and Physiology Part A: Physiology*, *47*(2), 521–28.

Tamaki, F. K., Padilha, M. H. P., Pimentel, A. C., Ribeiro, A. F. & Terra, W. R. (2012). "Properties and Secretory Mechanism of Musca Domestica Digestive Chymotrypsin and Its Relation with Drosophila Melanogaster Homologs." *Insect Biochemistry and Molecular Biology*, *42*(7), 482–90.

Tamaki, F. K., Pimentel, A. C., Dias, A. B., Cardoso, C., Ribeiro, A. F., Ferreira, C. & Terra, W. R. (2014). "Physiology of Digestion and the Molecular Characterization of the Major Digestive Enzymes from Periplaneta Americana." *Journal of Insect Physiology*, *70*, 22–35.

Tamaki, F. K. & Terra, W. R. (2015). "Molecular Insights into Mechanisms of Lepidopteran Serine Proteinase Resistance to Natural Plant Defenses." *Biochemical and Biophysical Research Communications*, *467*(4), 885–91.

Terra, W. R. (1988). "Physiology and Biochemistry of Insect Digestion: An Evolutionary Perspective." *Brazilian Journal of Medical and Biological Research*, *21*(4), 675-734.

Terra, W. R. (1990). "Evolution Of Digestive Systems Of Insects." *Annual Review of Entomology*, *35*(1), 181–200.

Terra, W. R. & Cristofoletti, P. T. (1996). "Midgut Proteinases in Three Divergent Species of Coleoptera." *Comparative Biochemistry and Physiology B Biochemistry and Molecular Biology*, *113*(4), 725–30.

Terra, W. R. & Ferreira, C. (1983). "Further Evidence That Enzymes Involved in the Final Stages of Digestion by Rhynchosciara Do Not Enter the Endoperitrophic Space." *Insect Biochemistry*, *13*(2), 143–50.

Terra, W. R. & Ferreira, C. (1994). "Insect Digestive Enzymes: Properties, Compartmentalization and Function." *Comparative Biochemistry and Physiology*, *109*(1), 1–62.

Terra, W. R. & Ferreira, C. (2005). "Biochemistry of Digestion." In *Biochemistry*, 171–224.

Terra, W. R. & Ferreira, C. (2012). "Biochemistry and Molecular Biology of Digestion." In *Insect Molecular Biology and Biochemistry*, 365–418, Elsevier B.V.

Terra, W. R., Ferreira, C. & Garcia, E. S. (1988). "Origin, Distribution, Properties and Functions of the Major Rhodnius-Prolixus Midgut Hydrolases." *Insect Biochemistry*, *18*(5), 423–34.

Tetreau, G., Stalinski, R., David, J. P. & Després, L. (2013). "Increase In Larval Gut Proteolytic Activities And Bti Resistance In The Dengue Fever Mosquito." *Archives of Insect Biochemistry and Physiology*, *82*(2), 71–83.

Tham, H. W., Balasubramaniam, V. R. M. T., Tejo, B. A., Ahmad, H. & Hassan, S. S. (2014). "CPB1 of Aedes Aegypti Interacts with DENV2 E Protein and Regulates Intracellular Viral Accumulation and Release from Midgut Cells." *Viruses*, *6*(12), 5028–46.

Thomas, A. W., Roseth, B. M. & Gooding, R. H. (1976). "Digestive Processes of Hematophagous Insects. X. Effects of Mammalian Sera upon Two Trypsins from Female Horse Flies and Deer Flies (Diptera: Tabanidae)." *Journal of Medical Entomology*, *13*(3), 341–46.

Tong, N. T., Imhoff, J. M., Lecoisey, A. & Keil, B. (1981). "Hypodermin A, a Trypsin-like Neutral Proteinase from the Insect Hypoderma Lineatum." *Biochimica et Biophysica Acta*, *658*, 209–19.

Upadhyay, S. K. & Chandrashekar, K. (2012). “Interaction of Salivary and Midgut Proteins of Helicoverpa Armigera with Soybean Trypsin Inhibitor.” *Protein Journal*, *31*(3), 259–64.

Valaitis, A. P. (1995). “Gypsy Moth Midgut Proteinases: Purification and Characterization of Luminal Trypsin, Elastase and the Brush Border Membrane Leucine Aminopeptidase.” *Insect Biochemistry and Molecular Biology*, *25*(1), 139–49.

Valdés-Rodríguez, S., Galván-Ramírez, J. P., Guerrero-Rangel, A. & Cedro-Tanda, A. (2015). “Multifunctional Amaranth Cystatin Inhibits Endogenous and Digestive Insect Cysteine Endopeptidases: A Potential Tool to Prevent Proteolysis and for the Control of Insect Pests.” *Biotechnology and Applied Biochemistry*, *62*(5), 634–41.

Venkat, V., Kumar, S. K., Sridevi, P., Muley, V. Y. & Chaitanya, R. K. (2017). “Cloning, Characterization and Transmission Blocking Potential of Midgut Carboxypeptidase A in Anopheles Stephensi.” *Acta Tropica*, *168*, 21–28.

Verma, S., Das, S., Mandal, A., Ansari, M. Y., Kumari, S., Mansuri, R., Kumar, A., et al. (2017). “Role of Inhibitors of Serine Peptidases in Protecting Leishmania Donovani against the Hydrolytic Peptidases of Sand Fly Midgut.” *Parasites and Vectors*, *10*(1), 1–16.

Vorster, J., Rasoolizadeh, A., Goulet, M. C., Cloutier, C., Sainsbury, F. & Michaud, D. (2015). “Positive Selection of Digestive Cys Proteases in Herbivorous Coleoptera.” *Insect Biochemistry and Molecular Biology*, *65*, 10–19.

Waniek, P. J., Araújo, C. A. C., Momoli, M. M., Azambuja, P., Jansen, A. M. & Genta, F. A. (2014). “Serine Carboxypeptidases of Triatoma Brasiliensis (Hemiptera, Reduviidae): Sequence Characterization, Expression Pattern and Activity Localization.” *Journal of Insect Physiology*, *63*(1), 9–20.

Wei, J., Zhang, M., Liang, G., Wu, K., Guo, Y., Ni, X. & Li, X. (2016). “APN1 Is a Functional Receptor of Cry1Ac but Not Cry2Ab in Helicoverpa Zea.” *Scientific Reports*, *6*, 1–8.

Whitworth, S. T., Blum, M. S. & Travis, J. (1998). “Proteolytic Enzymes from Larvae of the Fire Ant, Solenopsis Invicta: Isolation and

Characterization of Four Serine Endopeptidases." *Journal of Biological Chemistry*, *273*(23), 14430–34.

Whitworth, S. T., Kordula, T. & Travis, J. (1999). "Molecular Cloning of Soli E2: An Elastase-like Serine Proteinase from the Imported Red Fire Ant (Solenopsis Invicta)." *Insect Biochemistry and Molecular Biology*, *29*(3), 249–54.

Wielkopolan, B., Walczak, F., Podleśny, A., Nawrot, R. & Obrepalska-Steplowska, A. (2015). "Identification and Partial Characterization of Proteases in Larval Preparations of the Cereal Leaf Beetle (Oulema Melanopus, Chrysomelidae, Coleoptera)." *Archives of Insect Biochemistry and Physiology*, *88*(3), 192–202.

Wigglesworth, V. B. (1953). *The Principles of Insect Physiology*. Fifth. London: Methuen.

Yang, Y. J. & Davies, D. M. (1968). "Digestion, Emphasizing Trypsin Activity, in Adult Simuliids (Diptera) Fed Blood, Blood-Sucrose Mixtures, and Sucrose." *Journal of Insect Physiology*, *14*(2), 205–22.

Yao, J., Buschman, L. L., Oppert, B., Khajuria, C. & Zhu, K. Y. (2012). "Characterization of CDNAs Encoding Serine Proteases and Their Transcriptional Responses to Cry1Ab Protoxin in the Gut of Ostrinia Nubilalis Larvae." *PLoS ONE*, *7*(8), e44090.

Zeng, F., Zhu, Y. C. & Cohen, A. C. (2002). "Molecular Cloning and Partial Characterization of a Trypsin-like Protein in Salivary Glands of Lygus Hesperus (Hemiptera: Miridae)." *Insect Biochemistry and Molecular Biology*, *32*(4), 455–64.

Zhang, Ti., Coates, B. S., Wang, Y. Q., Wang, Y., Bai, S., Wang, Z. Y. & He, K. (2017). "Down-Regulation of Aminopeptidase N and ABC Transporter Subfamily G Transcripts in Cry1Ab and Cry1Ac Resistant Asian Corn Borer, Ostrinia Furnacalis (Lepidoptera: Crambidae)." *International Journal of Biological Sciences*, *13*(7), 835–51.

Zhao, B., Kokoza, V. A., Saha, T. T., Wang, S., Roy, S. & Raikhel, A. S. (2014). "Regulation of the Gut-Specific Carboxypeptidase: A Study Using the Binary Gal4/UAS System in the Mosquito Aedes Aegypti." *Insect Biochemistry and Molecular Biology*, *54*, 1–10.

Zhou, X., Fan, D. & Zhao, K. (2016). "Characterization of Trypsin-like and Chymotrypsin-like Serine Proteases from Midgut of Mythimna Separata Walker." *Archives of Insect Biochemistry and Physiology*, *92*(3), 173–91.

Zhu, Y. C. & Baker, J. E. (1999). "Characterization of Midgut Trypsin-like Enzymes and Three Trypsinogen CDNAs from the Lesser Grain Borer, Rhyzopertha Dominica (Coleoptera: Bostrichidae)." *Insect Biochemistry and Molecular Biology*, *29*(12), 1053–63.

Zibaee, A., Hoda, H. & Mahmoud, F. D. (2012). "Role of Proteases in Extra-Oral Digestion of a Predatory Bug, *Andrallus Spinidens*." *Journal of Insect Science*, *12*(51), 1–17.

Zinkler, D. & Polzer, M. (1992). "Identification and Characterization of Digestive Proteinases from the Firebrat Thermobia Domestica." *Comparative Biochemistry and Physiology*, *103*(3), 669–73.

Zulkifli, A. N., Zakeri, H. A. & Azmi, W. A. (2018). "Food Consumption, Developmental Time, and Protein Profile of the Digestive System of the Red Palm Weevil, Rhynchophorus Ferrugineus (Coleptera: Dryophthoridae) Larvae Reared on Three Different Diets." *Journal of Insect Science*, *18*(5), 1–7.

In: Proteases: Functions, Mechanisms and Uses ISBN: 978-1-53615-854-0
Editor: Nelson Pérez Guerra

Chapter 3

MARINE PROTEASES

Arisaí C. Hernández-Sámano[1], Raquel García-Barrientos[2], Xóchitl Guzmán García[3] and Isabel Guerrero Legarreta[3,*]

[1]Department of Cellular and Molecular Neurobiology, Universidad Nacional Autónoma de México-Juriquilla Campus, Querétaro, Mexico

[2]Universidad Politécnica de Tlaxcala, San Pedro Xalcaltzinco, Tlaxcala, Mexico

[3]Biological and Health Sciences Division, Universidad Autónoma Metropolitana-Iztapalapa Campus, Mexico City

ABSTRACT

Organisms inhabiting the marine environments are excellent sources of novel enzymes, which retain their integrity and function under extreme or different reaction conditions. Proteolytic enzymes, or proteases, are a group of enzymes able to hydrolyze protein peptide bonds. These are classified by several criteria. By their optimal pH, they are neutral, acidic or alkaline. With respect to their active center, enzymes are divided into

[*] Corresponding Author's E-mail: isabel_guerrero_legarreta@yahoo.com.

serine, cysteine, aspartyl, and metallo-proteases. Marine proteases have unique properties, mainly high activity at low-temperatures, thermal stability and pH adaptation. Novel enzymes with particular characteristics or functions are valuable in industrial bioprocessing. In this respect, functions and mechanisms of marine proteases are similar to proteases from microorganisms, plants, and warm-blood animals, although their application can be in wider processing conditions, the action mechanism of specific marine proteases are studied at present by several *in silico* analyses. Marine protease applications are extensive. In aquaculture, in developing diets to several species. In the food industry, proteases are widely used for meat tenderization, fat emulsifying, foaming, beverage clarification, debittering, texturization, deproteinization and flavor production. In therapeutics, biopeptides produced by hydrolyzing specific proteins act as antioxidants, anticoagulants, antihypertensive, antitumor, antinflammation and anticancer, among others. Collagen obtained from marine sources using marine enzymes is also used in several industries, such as cosmetics, foods and medicine. Additionally, collagen hydrolysates are also bioactive (antioxidant, antimicrobial, antifatigue, anticancer, with mineral and hormonal regulating properties and angiotensin I converting enzyme or ACE). In addition, bioactive peptides and enzymes are also obtained from fish and seafood wastes such as shells, viscera, heads, skins, bones, fins and trimmings, helping to alleviate a serious pollution problem, as around 75% part of fish is discarded as a waste.

Keywords: marine protease, proteolytic enzyme, fish and shellfish

1. INTRODUCTION

The marine environment represents a vast resource for new bioactive natural products, with application in basic research, biomedical sciences, development of therapeutics, and a wide variety of industrial processes. Marine organisms produce unique molecules due to conditions that significantly differ from the land environment, including aggressive, demanding and highly competitive habitats, leading to the production of potent active molecules (Prashant and Padma, 2018). This environment is a niche of the largest, diversified genetic pool of material with an enormous potential of a wide variety of enzymes, including proteases. About 75% of industrial enzymes are proteases, representing about 60% of total worldwide sale of enzymes. A number of industries make use of proteases, including detergents, leather processing, silver recovery, medical purposes, food and feed processing, chemicals and waste treatment. While microorganisms are the source of most commercialized peptidases, marine eukaryotic organisms represent a significant stock of valuable peptidases with specific characteristics and potential applications and commercialization (Homaeia et al., 2016). Marine proteases considerably differ from similar molecules in internal or external organs of invertebrates and vertebrates. Some of these distinctive features include higher catalytic efficiency at low reaction temperatures, lower thermostability, stability at low temperature and substantial catalytic activity/stability at neutral to alkaline pH (Haard and Simpson, 2000). Numerous peptidases from marine animals had been characterized, showing stability at extreme pH and temperatures, as well as in the presence of organic solvents, metal ions and proteolytic agents (Homaeia et al., 2016).

2. Definition and Classification of Marine Proteases

Proteases are a large group of hydrolytic enzymes (catalyzing chemical bond disruption and adding water molecules) able to cleave peptide bonds in protein molecules and, subsequently, producing small peptides and amino acids (Prashant and Padma, 2018).

Proteases are divided into two groups, according to their mode of action: endopeptidases and exopeptidases. Endopeptidases hydrolyze peptide bonds located in mid positions of polypeptide chains, while exopeptidases remove terminal amino acid residues of polypeptide chains. Alternatively, proteases are classified by their optimal activity pH as neutral, acidic or alkaline. With respect to their active center, marine proteases (EC 3.4.) are classified into serine (EC 3.4.21.), cysteine (EC 3.4.22.), aspartyl (EC 3.4.23.) and metallo- (EC 3.4.24.) proteases (Homaeia et al., 2016).

2.1. Serine Proteases

This is a group of endoproteases with one serine residue as catalytic site. It is characterized by the presence of a serine residue, an imidazole group and an aspartyl carboxyl group in the catalytic site. Two distinct families are included, according to their structural homology: trypsin and subtilisin. Trypsin is the largest enzyme family found in both mammalian and bacterial members. Conversely, subtilisin family is only found in bacteria (Sriket et al., 2014). The three major serine proteases purified and characterized so far from digestive glands of marine animals are trypsin, chymotrypsin, and elastase (Haard and Simpson, 2000). These proteases are inhibited by N-tosyl-L-phenylalaninechloromethyl ketone (TPCK) and phenyl methyl sulfonyl fluoride (PMSF).

2.1.1. Trypsin

Trypsin (EC 3.4.21.4) has very narrow specificity for peptide bonds in the carboxyl side of arginine and lysine. Dornelles et al. (2018) purified a trypsin-like protease from viscera of mussel *Mytella charruana*. The enzyme, with pI 4.1, showed one single polypeptide band of 83.1 kDa when analyzed by SDS-PAGE. Its optimal activity is at 40°C, and is stabile in a wide pH range, 3.0 at 9.0. Km was 4.28 ± 0.34 mM in N–benzoyl–DL-arginyl-ρ-nitroanilide substrate, whereas V_{max} was 0.056 ± 0.001 nmol/ min. The enzyme hydrolyzed casein; the hydrolysate inhibited the growth of *Escherichia coli, Micrococcus luteus, Bacillus subtilis* and *Klebsiella pneumoniae* at a minimal inhibitory concentration of 5.0 µg/mL. In addition, *M. charruana* is an indicator of marine pollution and is an aquatic model for monitoring bioaccumulation of polyaromatic hydrocarbons in estuarine ecosystems (Maioli et al., 2010).

2.1.2. Chymotrypsin

Chymotrypsin endopeptidase (EC 3.4.21.1) has broader specificity than trypsin. It cleaves peptide bonds involving amino acids with bulky side chains and nonpolar amino acids such as tyrosine, phenylalanine, tryptophan and leucine (Haard and Simpson, 2000). Chymotrypsin, secreted by fish pancreatic tissues, has higher specific activity than bovine chymotrypsin (Zhou et al., 2011). This protease has been isolated and characterized from several fish species, such as freshwater and marine fish, such as Crucian carp (*Carassius auratus*), Monterey sardine (*Sardinops sagax caeruleus*), Japanese sea bass (*Lateolabrax japonicus*) and striped seabream (*Lithognathus mormyrus*) (Márquez-Ríos et al., 2016). The same authors reported isolating chymotrypsin from jumbo squid hepatopancreas. Its molecular weight is 31 kDa, peak activity at pH 7 and 65°C in Suc-Ala-Ala-Pro-Phe-p-nitroanilide (SAAPNA) substrate, and identified using TPCK and PMSF as specific inhibitors, showing residual activities of 6 and 0%, respectively. The peak activity was observed between pH 4.0 and 8.0. The purified enzyme showed moderate *in vitro* activity using muscle collagen as substrate.

2.1.3. Elastase

Elastase (EC 3.4.21.11) exhibits preferential specificity for alanine, valine, and glycine. Elastases were obtained from Atlantic cod (*Gadus morhua*) (Ásgeirsson and Bjarnason, 1993), king crab (*Paralithodes camtschatica*) (Sakharov et al., 1993) and rainbow trout (*Oncorhynchus mykiss*) (Bassompierre et al., 1993). Jayaraman et al. (2006) investigated metal binding and conformational stability of a psychrophilic elastase (ACE) from Atlantic cod (*Gadus morhua*). These authors reported that Ca^{2+} chelation was fundamental for maintaining the biologically active conformation and for thermal stability of the enzyme. However, presence of metal ions, such as Zn^{2+}, Fe^{3+} and Cu^{2+} inhibit the hydrolytic activity, also occurring for EDTA chelation. ACE exhibited a two-state unfolding pattern in both, acidic and basic conditions, with transition midpoint at pH 4.08 and 10.29, respectively. Stability characteristics of ACE have been compared with data for mesophilic porcine pancreatic elastase and a possible ACE mechanism for low temperature adaptation has been proposed.

2.1.4. Cathepsin

The name cathepsin is derived from the Greek *kathepsein* (to digest). Cathepsins are categorized into multiple families: serine, cysteine or aspartyl protease (Patel et al., 2018). Cathepsins A (EC 3.4.21.) and G (EC 3.4.21.20) are identified as serine proteases (Saikhedkar et al., 2015; Siklos et al., 2015). Cathepsin A (CTSA) is classified as exopeptidase and part of the papain family of serine protease. CTSA was identified in turbot (*Scophthalmus maximus*) (SmCTSA) and characterized from its transcriptome. The recombinant SmCTSA protein was successfully purified and its strong binding ability with three microbial ligands was described. CTSA is a multifunctional protease with catalytic and protective functions (Fu et al., 2018).

2.2. Cysteine Proteases

Thiol or cysteine proteases (EC 3.4.22.) are a group of endoproteases containing a Cysteine-Histidine-Aspartate (Cys-His-Asp) triad at the active site. A histidine residue in the active site acts as proton donor and enhances the nucleophilicity of the cysteine residue (Verma et al., 2016). Thiol proteases are inhibited by heavy metal ions and their derivatives, as well as by alkylating and oxidizing agents (Haard and Simpson, 1994, 2000). In 1937, papain was the first cysteine protease isolated and characterized from *Carica papaya*. Papain and cathepsin belong to the most abundant family of the cysteine proteases.

2.2.1. Cathepsin

As pointed out before, cathepsins are classified into different families. Cathepsin B, C, F, H, K, L, O, S, V/L2, W and Z/X (EC 3.4.22.) were categorized as lysosomal cysteine proteases (Saikhedkar et al., 2015; Siklos et al., 2015). Cathepsins are involved in cephalopod intracellular digestion at acidic pH environment in the digestive gland, a highly active anatomical region. In *Dosidicus gigas* and *Octopus maya*, protease activity of the digestive gland has been attributed to intracellular cathepsins. In the case of

D. gigas the presence of cathepsin L (EC 3.4.22.15) has been reported by Cárdenas-López and Haard (2009). Godahewa et al. (2017) identified a CTSZ homolog of cathepsin Z, a lysosomal cysteine protease (AbCTSZ) from disk abalone (*Haliotis discus*), characterize at molecular, genomic, and transcriptional levels. AbCTSZ encodes a protein with 318 amino acids and a molecular mass of 36 kDa.

2.2.2. Calpain

Calpains (EC 3.4.22.52-54) are cysteine proteases, needing calcium ions for activation. Two types of calpain have been isolated, differing in their calcium requirement: μ-calpain and m-calpain. The peak pH is neutral to weakly alkaline (pH ~7.5). Calpains are primarily active within the first 24 hours *postmortem* (Sriket, 2014), activity causes limited hydrolysis of myofibrils during initial *postmortem* stages, whereas cathepsins have the capacity to breakdown actin and myosin at later *postmortem* storage. m-calpains are present in striated muscle of all marine species, such as tilapia, salmon and trout (Ahmed et al., 2015). Ge et al. (2018) investigated the differential role of calpain and cathepsin B, L, D in degradation and disassembly of myofibrillar proteins of grass carp (*Ctenopharyngodon idella*). Striated muscle samples were incubated with the three enzymes. The results showed that calpains and cathepsin B, and calpain and cathepsin L act in a stepwise and complementary manner to synergistically dissociate and degrade myofibrillar proteins.

2.3. Aspartyl Proteases

Aspartic proteases, aspartyl proteases (EC 3.4.23.), or carboxyl proteases are a group of endoproteases characterized by high activity and stability at low pH. Their catalytic sites are composed of the carboxyl groups of two aspartic acid residues (Asp32 and Asp215) located between two lobes. Three aspartyl proteases have been isolated and characterized from the stomach of marine animals: pepsin, chymosin and gastricsin (Haard and

Simpson, 2000) and, recently, cathepsins. These enzymes are inhibited by Pepstatin A.

2.3.1. Pepsin

Pepsin (EC 3.4.23.1) has specificity for aromatic amino acids (phenylalanine, tyrosine, and tryptophan). Pepsinogen and pepsin have been extensively studied in freshwater and marine fish species, such as Antarctic rock cod (*Trematomus bernacchii*) (Brier et al., 2007) and pectoral rattail (*Coryphaenoides pectoralis*) (Klomklao et al., 2007). However, there is scarce information about tropical fish digestive enzymes (Castillo-Yañez et al., 2004). Acevedo et al. (2018) obtained acid proteases by salting-out of shad (*Prochilodus lineatus*) stomach mucosa. This single step process produced an enzyme extract purified 1.8-fold of the crude extract, 45.1% recovery of initial proteolytic activity. Shad proteases exhibited peak activity at 45°C, pH 2.0, are stable at pH 2.0 to 5.0, retaining more than 70% of the initial activity after incubation at pH 7.0 for 2 hours. Fish extracts were unstable above 45°C; as compared to the acid extraction method, the use of the shad enzymatic extract increased 1.7 times the collagen extraction yield.

2.3.2. Gastricsin

Gastricsin (EC 3.4.23.3) are aspartyl proteases with similar enzymatic and chemical properties of pepsin. However, they differ from pepsins in structure and certain catalytic properties. For example, gastricsins are active at extremely low pH, and usually its peak activity is at pH 2 to 3 in hemoglobin substrates (De Luca et al., 2009). These enzymes have specific activity two-fold higher in peptides, with preference for Tyr at P1 position (Carginale et al., 2004). Some gastricsins were extracted from salmon (*Salmo salar*) and hake (*Merluccius gayi*) gastric juice (Haard and Simpson, 2000). A gastricsin (named pepsin C) was isolated from the gastric mucosa of Antarctic notothenioid rock cod (*Trematomus bernacchii*) (De Luca et al., 2009).

2.3.3. Chymosin

Chymosin (EC 3.4.23.4), formerly known as rennin, has specificity for aromatic amino acids: phenylalanine, tyrosine and tryptophan, in a similar way to pepsin. Chymosin and chymosin-like enzymatic activities were reported in carp (Shahidi and Kamil, 2001). Chymosin differ from pepsin in some properties. For example, harp seal chymosins properties include a relatively higher milk clotting/proteolytic activity ratio (Rossano et al., 2011), peak pH 2.2-3.5 of hemoglobin hydrolysis, as compared to pH 2.0 for mammalian pepsins. Other differences are chymosin inability to inactivate ribonuclease, low activity on N-acetyl-L-phenylalanyl-3,5-diiodo-L-tyrosine substrates, non-stabile in 6.0 M urea and zymogen molecular weight of 33.8 kDa (Haard and Simpson, 2000).

2.3.4. Cathepsin

Cathepsins are categorized into three families: aspartyl, cysteine or serine protease (Patel et al., 2018). Cathepsins D (EC 3.4.23.5) and E (EC 3.4.23.34) are aspartate proteases (Saikhedkar et al., 2015; Siklos et al., 2015). Aoki (2000) found cathepsin D in red or white muscles of 24 animal species; no differences were found between red- and white-flesh fish, or freshwater fish. Cathepsin D1, isolated from American lobster (*Homarus americanus*) gastric fluid, was reported being an aspartic peptidase, highly active at 5 to 50°C and pH 2.5 to 5.5. Non-ionic detergents, such as Tween-20 and Triton X-100, stabilize the peptidase activity. Ethanol, methanol and isopropanol [5 to 15% (v/v)] increased enzyme activity up to 80%. This enzyme is active up to 2.5 M urea and resistant to proteolysis by papain and renin. At 5 to 35°C the catalytic efficiency is 20-fold higher than bovine and porcine similar enzymes. In fact, lobster aspartic peptidase is named a cold-adapted enzyme (Rodríguez-Siordia et al., 2018). These characteristics are important in pharmaceutical, cosmetic and food applications. Another cathepsin, probably cathepsin D, an acid protease obtained from *Octopus maya*, was found to play a basic role as octopus extracellular digestive protease (Martínez et al., 2011, 2012).

2.3.5. Nothepsin

De Luca et al. (2009) named newly identified aspartic proteases "Nothepsins" (EC 3.4.23.). The authors identified two novel aspartic proteases from two Antarctic fish species liver, the red-blooded *Trematomus bernacchii* and the haemoglobinless icefish *Chionodraco hamatus*. Alternatively, the proteinase found in Antarctic fish may be a modifications of an ancestral aspartic proteinase gene. The name "Nothepsin" may be appropriate for this enzyme.

2.4. Metalloproteases

Metalloproteases (EC 3.4.24.) are hydrolytic enzymes, activity depends on the presence of bound divalent cations. Chemical modification studies suggest that there is, at least, one tyrosyl and one imidazole residue associated to the catalytic site. These enzymes are inhibited by chelating agents such as 1,10-phenanthroline, EDTA and, sometimes, by simple dialysis. Most of the known metalloproteases are exopeptidases (Haard and Simpson, 2000).

2.4.1. Collagenase

Collagenases (EC 3.4.24.) degrade the various types of collagens, the most abundant constituents of the extracellular matrix in fibrous structural proteins in higher organisms. Collagens are mostly found in fibrous tissues such as skin, ligaments and tendons in the form of elongated fibrils. These proteins are also abundant in cornea, blood vessels, bone, cartilage, intervertebral disc and gut. In mammals, these proteins represent over 30% of total protein in the animal body. Due to the rigid triple helical structure of collagens, they are resistant to the action of most proteases but can be readily cleaved by site-specific action of collagenases (Prashant and Padma, 2018). Krishnamurthy and Belur (2018) purified and characterized a fibrinolytic serine metalloprotease from marine *Serratia marcescens* subsp. *sakuensis* (KU296189.1). The purified 43 kDa enzyme (1033 U/mg) peak activity is at pH 7 and 55°C; *in vitro* half-life at 37°C was 19 h. Km and V_{max} of the purified enzyme in fibrin substrate is 0.66 mg/mL and 158.73 U/mL, respectively; Kcat and catalytic efficiency is 12.21 min^{-1} and 18.32 mL/(mg min), respectively. This fibrinolytic enzyme does not show any proteolytic activity towards blood plasma proteins, such as haemoglobin, γ-globulins and transferrin. *In vitro* studies reported that the fibrinolytic enzyme displayed 38% clot lysis 3 h, higher than clot lysis by streptokinase and heparin. Seven peptide sequences were obtained by LC-MS/MSTOF analysis, only four sequences showed 67% similarity at amino acid level to other proteases.

3. Characteristics Development of Marine Proteases by Adaptation Marine Organisms to Environmental Conditions

Proteins are biochemical engines, therefore related to functional processes of how a given organism responds to changing environments. The proteome is a dynamic and complex network responding in many ways (Tomanek, 2014). In all enzymes, external factors such as temperature, pH and media composition determine activity, catalytic efficiency, stability and proper functioning. However, marine proteases differ to their land counterparts in sensitivity to inhibitors, catalytic efficiency or substrate affinity reported for marine proteases. These two groups of enzymes considerably differ in their location in internal or external organs of invertebrates and vertebrates. For example, trypsin (the most abundant protease) appeared early in evolution, occurring in the digestive systems of invertebrates in all phyla. Even though, trypsin shows differences in invertebrates, depending on feeding regimes and habitats. Life style, survival ability and environment cause these differences (Muhlia-Almazán et al., 2008).

4. Proteolytic Extremozymes from Marine Microorganisms

Extremely hot terrestrial and marine environments are hostile for most living organisms, but these niches represent excellent settings to specialized microorganisms (Elleuche et al., 2015). Marine microorganisms (Bacteria and Archaea) able to survive in diverse conditions, such as extreme temperature, pH, pressure and salt concentration are known as extremophiles. They produce biocatalysts, called extremozymes, active and stable at extreme conditions (Suriya et al., 2016). Among these organisms are polychaeta, a paraphyletic class of annelid worms, generally living in the marine environment. These robust and widespread organisms can live in a wide variety of conditions, from the abyssal plain –the coldest ocean temperature– to hydrothermal vents at extremely high temperature. *Capitella sp. I* and *Perinereis nuntia brevicirris* are two polychaete living in different coastal habitats. *Capitella* sp. I is found in sediments beneath fish farms, while *P. nuntia brevicirris* inhabits estuaries. The proteases extracted from these two polychatae show activity adapted to their habitat. A novel protease isolated from polychaete *Periserrula leucophryna* is highly stabile to extreme pH and temperature, detergents, oxidizing agents, heavy metals and some solvents, make it suitable for industrial applications (Homaei et al., 2016).

5. Functions of Marine Proteases

Proteolytic enzymes are crucial for a variety of biological processes in organisms ranging from lower (virus, bacteria, and parasite) to the higher organisms (Verma et al., 2016). Proteases participate in essential processes such as food digestion, cell growth, morphogenesis, development, inflammation, and activation of zymogens and prohormones. Degradomics is a sub-discipline of biology encompassing all the genomic and proteomic approaches devoted protease study. It is focused on elucidating proteases present at a given time in a particular medium (which may be cross-referenced with the physiological state of a specific specimen), including information on a enzyme-substrate and/or enzyme-inhibitor (Tavano et al., 2018). A degradome is a diversity of expressed proteases in an organism, tissue, or cell at a given time.

5.1. Serine Proteases

Serine endoproteases are divide into two superfamilies that independently evolved similar catalytic mechanisms: trypsin superfamily (trypsins and chymotrypsins) and subtilysine superfamily (present in most bacteria and fungi). Invertebrate proteolytic enzymes have been widely studied, special attention is focused in the trypsin group because they take part in a number of physiological processes, playing a major and better-understood key role in hydrolyzing food proteins (Muhlia-Almazán et al., 2008). Chymotrypsin-like proteases are involved in many critical physiological processes, including digestion, hemostasis, apoptosis, signal transduction, reproduction, and immune response (Hedstrom, 2002). For example, alkaline enzymes (mainly trypsin and chymotrypsin) in octopus species *O. maya* and *O. vulgaris* play an important role in extracellular digestion of preys (Ibarra-García et al., 2018).

5.2. Cysteine Proteases

The cysteine protease family includes several lysosomal cathepsins, cytoplasmic calpains (calcium-activated) as well as several parasitic proteases. Lysosomal cysteine proteases, generally known as cathepsins, play an important role in many physiological processes, such as protein degradation (Sriket, 2014). Cathepsins, like other enzymes, are vital for normal physiological functions, such as digestion, blood coagulation, bone resorption, ion channel activity, innate immunity, complement activation, apoptosis, vesicular trafficking, autophagy, angiogenesis, proliferation, and metastasis, among others (Patel et al., 2018). Thus, cathepsins play a fundamental in maintaining homeostasis. Most lysosomal cysteine cathepsins belong to the C1 peptidase family, and also known as papain family. Cathepsins also include three groups: cysteine proteases (cathepsins B, C, F, H, K, L, O, S, W and X or Z), aspartyl proteases (cathepsins D and E) and serine proteases (cathepsins A and G), based on amino acid motifs (Fu et al., 2018).

5.3. Aspartyl Proteases

A number of cell and tissue produce aspartyl proteases, most of them belong to the pepsin family that includes digestive enzymes such as pepsin and chymosin, as well as lysosomal cathepsin D and processing enzymes. They play important roles in protein digestion at low pH (Sriket, 2014).

5.4. Metalloproteases

Metalloproteases include enzymes from a variety of origins, such as collagenases in higher organisms, toxins in snake venoms, and thermolysin in bacteria. These enzymes contain a catalytically active zinc atom. In some cases, other metal, such as cobalt or nickel, replaces the zinc without losing activity. Matrix metalloproteases (MMPs) are a family of zinc endopeptidases responsible for collagen degradation in extracellular fluids (Sriket, 2014). Collagenases are involved in normal growth and development, and in tissue repair by precise regulation of collagen catabolism and synthesis (Sivakumar et al., 1999). The combination of structural and functional studies of metalloproteases revealed the presence of other members of this family, important to hormone processing, immune/inflammatory activity and blood-clotting functions (Auld, 2013).

6. Mechanism of Marine Proteases

All proteases must overcome three obstacles to hydrolyze a peptide bond: (a) amide bonds are very stable due to electron donation of amide nitrogen to carbonyl. Proteases usually activate an amide bond via the interaction of carbonyl oxygen with a general acid, and may also distort the peptide bond to disrupt resonance stabilization; (b) water is a poor nucleophile; proteases always activate water, usually via a general base; and (c) amines are poor leaving groups; proteases protonate the amine prior to expulsion. Thus, the catalysis of peptide hydrolysis involves activation of the carbonyl via a general acid, activation of water with a general base, and protonation of the amine-leaving group (Figure 1) (Hedstrom, 2002).

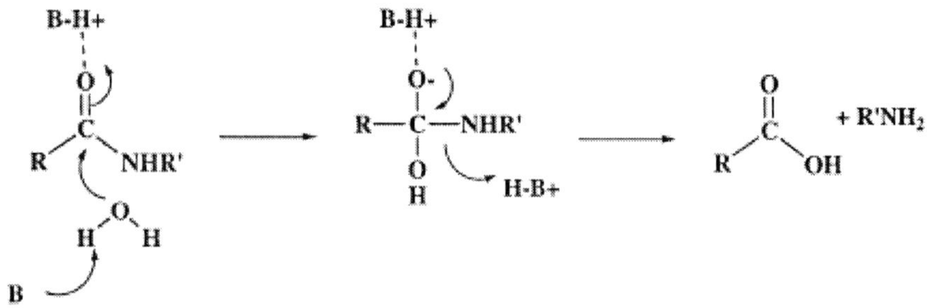

Figure 1. Catalysis of peptide hydrolysis (Hedstrom, 2002).

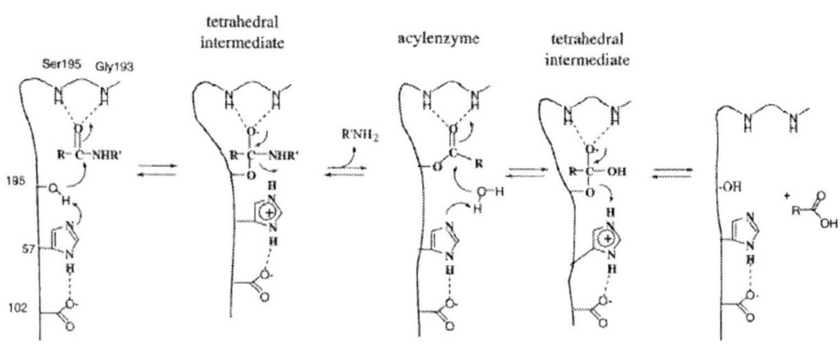

Figure 2. The generally accepted chemical mechanism of serine protease catalysis for serine proteases (Hedstrom, 2002).

6.1. Serine Proteases

Three steps are involved in the serine proteases mechanism (Haard and Simpson, 2000): Step 1, the serine OH group performs nucleophilic attack; Step 2, the imidazole ring of histidine transfers protons; Step 3, slightly positively charged NH groups fix and stabilize the substrate's carbonyl oxygen atom. Chymotrypsin has 245 residues arranged in two six-stranded beta barrels. The structure includes catalytic, substrate recognition and zymogen activation domains, common to all chymotrypsin-like serine proteases. The accepted mechanism for chymotrypsin-like serine proteases considers that the acylation half of the reaction, Ser195 attacks the carbonyl of the peptide substrate, assisted by His57 acting as a general base, to yield a tetrahedral intermediate. The resulting His57-H^+ is stabilized by the hydrogen bond to Asp102. The interaction between the oxyanion hole and the NHs main chain NH stabilizes the tetrahedral intermediate that, when collapsing assisted by His57-H^+ acting as a general acid, yields an acylenzyme intermediate (Figure 2) (Hedstrom, 2002).

6.2. Cysteine Proteases

Cysteine proteases contain a Cys-His-Asp triad at the active site. A histidine residue, present in the active site, acts as proton donor and enhance the nucleophilicity of the cysteine residue. Te nucleophilic cysteine residue attacks the carbon of a reactive peptide bond, producing the first tetrahedral thioester intermediate in the reaction with the release of an amine or amino terminus fragment of the substrate. Hydrogen bondings between the substrate oxyanion and a conserved glutamine residue stabilizes the intermediate. Subsequently, the thioester bond is hydrolyzed to produce a carboxylic acid moiety from the remaining substrate fragment (Verma et al., 2016).

6.3. Aspartyl Proteases

The number of structures in a variety of organisms depend on the occurrence of aspartic peptidases. The structure of an aspartic peptidase is described in several ways. On one hand, these enzymes are two-domain proteins where a gene duplication event has repeated the sequence. Studies of different inhibitors bound, has revealed that the two halves of the enzyme act as independent folding units that move relative to each other. Binding of the natural product pepstatin (inhibitor) to a wide variety of aspartic peptidases used as a model for the tetrahedral intermediate in the catalytic reaction (Dunn, 2002).

The chemical mechanism proposed for all aspartic peptidases (Figure 3) accounts for proton transfers to the substrate. However, intermediate FT (Step 3, Figure 3) leaves the carbonyl oxygen of the substrate in a negative charge without hydrogen bonding to stabilize this high-energy form. In the catalytic mechanism proposed by Northrop (Dunn, 2002), E species is the free enzyme poised for catalysis. Step 1 is the binding of substrate to form a loose complex (ES). Step 2 is the closing of the flap down upon the substrate to squeeze all components into the correct geometry and distances for the catalytic process to begin (ES). Step 3 includes the removal of a proton from the bound water molecule to stimulate attack on the carbonyl carbon (FT). Step 4 involves a proton transfer to the nitrogen of the peptide bond (GZ). Step 5 is the bond cleavage event (FPQ). Step 6 is the opening of the flap to free the products (FPQ) and step 7 is release of the products (F). Step 8 includes a loss of one proton (G) and step 9 involves binding of a new water molecule and re-formation of the low-barrier hydrogen bond (E) (Dunn, 2002).

6.4. Metalloproteases

Many metalloproteases contain the sequence of His-Glu-Xaa-Xaa-His (HGXXH), which provides two histidine ligands for the zinc, whereas the third ligand is either a glutamic acid (thermolysin, neprilysin, alanyl aminopeptidase) or a histidine (astacin). The catalytic mechanism leads to the formation of a non-covalent tetrahedral intermediate after the attack of a zinc-bound water molecule on the carbonyl group of the scissile bond. The intermediate is further decomposed by transfer of the glutamic acid proton to the leaving group (Sriket, 2014).

Three types of distinct zinc-binding motifs emerge from X-ray diffraction and nuclear magnetic resonance (NMR) analyses: 'catalytic', 'cocatalytic', and 'structural' sites. In zinc metalloenzymes, the most common amino acids that supply ligands to these sites are His, Glu, Asp, and Cys. Thus far, Cys is a ligand in only one metalloprotease (Auld, 2013). Collagenases are zinc-containing matrix metalloproteases (endopeptidases), that cleave collagen molecules at specific Gly-Ile bonds (a 1 of type I collagen) and Gly-Leu (a 2 of type I collagen) bonds, at approximately three quarters the distance from the amino terminus of the molecule (Sivakumar et al., 1999).

7. *In Silico* Analysis of Proteases

Bioinformatics, also known as *in silico* (computer simulation) analysis, involves computational methods applied to manage, curate, and interpret information related to biological systems. This field combines computer science, biology, mathematics, and statistical techniques to analyze and interpret biological data (Tu et al., 2018). Currently, bioinformatics methods are used for identifying, characterizing, elaborating bioactive mechanisms of, and producing food-derived bioactive peptides. These *in silico* methods include databases, online tools, and software. Table 2 lists some of these methods. Notably, among them, one tool may possess multiple functions.

Table 1. Properties of several marine and mammalian proteases (modified from Haard and Simpson, 2000)

Enzyme	Source	Biological Temp. (°C)	Molecular Weight (kDa)	pH optimum	Temp. optimum (°C)	K_m (mM)	V_{max}
Acid proteases							
Pepsin	Atlantic cod	0 – 5	35.5	3.5	40	0.05	
	Capelin	(-1) – 4	23.0; 7.0	3.7; 2.5	38; 43		
	Greenland cod	(-2) – 2	36.4	3.5	30	0.86	
	Polar cod	(-1.5) – 2	40.0; 42.0	2.0	37	0.06;0.33	31;556
	Sardine	10	33.0; 37.0	2.0; 4.0	40; 55		
	Porcine	37	32.0–35.0	2.0	47	0.04	
Chymosin Gastricin	Hake		32.3; 33.9	3.0			
Serine proteases							
Trypsin	Atlantic cod	0-5	224.2; 24.8	7.5-8.0	40	1.42	0.25
	Capelin	(-1) - 4	28.0	7.5-8.0	45	0.73; 1.38	
	Cunner	>0.5	24.0	8.5	45	0.73	0.18
	Greenland cod	(-2) – 2	23.5	7.5	35	1.67	0.21
	Mullet	15	24.0	8.0	55	0.49	400
	Sardine	10	22.9-28.8	10.0	45		
	Bovine	37	23.0	8.2	45	1.02	0.03

Table 1. (Continued)

Enzyme	Source	Biological Temp. (ºC)	Molecular Weight (kDa)	pH optimum	Temp. optimum (ºC)	K_m (mM)	V_{max}
Chymotrypsin	Atlantic cod	0-5	26.0	7.8	52	0.14; 0.20	207;214
	Dogfish			7.0; 8.0		3.5; 0.27	
	Rainbow trout	15	28.2-28.8	9.0	55	0.04; 0.02	2.24; 2.52
	Bovine	37		7.2	57	0.053	0.85
Elastasa	Atlantic cod	0-5	28.0	7.5-9.0	43	1.96	38.5
	Porcine	37		7.5-9.0	58	2.42	11.0
Thiol proteases							
Cathepsin B	Mussel		21.0	3.0-4.5			

Table 2. *In silico* methods contains some database, online tools, and software

Category	Name	Website	Function
Protein database	NCBI Protein database	http://www.ncbi.nlm.nih-gov/	Basic sequence information for proteins
	UniPeotKB	http://uniprot.org/	Basic sequence and structural information for proteins
	BIOPEP	http://www.uwm.edu.pl/biochemica/index.php/en/biopep	Protein sequence database
In silico digestion tools	PeptideCutter	http://web.expasy.org/peptide_cutter/	Server for predicting potential cleavage sites cleaved by proteases or chemical in a given protein sequence
	BIOPEP	http://www.uwm.edu.pl/biochemica/index.php/en/biopep	Server for predicting potential cleavage sites cleaved by proteases in a given protein sequence
	Enzyme Predictor	http://bioware.ucd.ie/-enzpred/Enzpred.php	Tool to evaluate the evidence for which enzymes are most likely to have cleaved a sample containing peptides from hydrolyzed proteins
Bioactive peptide database	BIOPEP	http://www.uwm.edu.pl/biochemica/index.php/en/biopep	Bioactive peptide database
	BitterDB	http://bitterdb.agri.huji.ac.il/bitterdb/	Bitter compounds database
	EROP-Moscow database	http://erop.inbi.ras.ru	Database of biologically active peptides
	APD	http://aps.unmc.edu/AP/main.htm/	Several kinds of bioactive peptide database Mainly focused on antimicrobial peptides
	PeptideDB	http://www.peptides.be/	Biologically active peptide database
	PepBank	http://pepbank.mgh.harvard.edu/	Biologically active peptide database providing searching program for fragments with sequence similar to the peptides in the database
	AHTPDB	http://crdd.osdd.net/raghava/ahtpdb	Antihypertensive peptide database
Potential bioactive prediction	BIOPEP	http://www.uwm.edu.pl/biochemia/index.php/en/biopep	Tool for the evaluation of proteins as the precursors of bioactive peptides
	PeptideRanker	http://bioware.ucd.ie/-compass/biowarewb/Server_pages/peptideranker.php	Server for the prediction of bioactive peptides

Table 2. (Continued)

Category	Name	Website	Function
	AntiBP2	http://crdd.osdd.net/raghava/anibp2/	Predicting the antibacterial peptides in a protein sequence
Allergenicity/toxicity prediction/ analyzing	AlgPred	http://crdd.net/raghava//antibp2/	Predicting allergenic proteins and peptides
	BIOPREP	http://www.uwm.edu.pl/biochemia/index.php/en/biopep	Allergenic protein database
	ToxinPred	http://crdd.osdd.net/raghava//toxinpred/	Predicting toxicity of peptides
Physicochemical characteristics prediction	Expasy-Compute pI/Mw	http://web.expasy.org/compute.pi/	Tool to compute the theoretical pI (isoelectric point) and Mw (molecular weight)
	Protparam	http://web.expasy.org/protparam/	Tool to compute gran average of hydropathicity (GRAVY) and instability index
	PepDraw	http://www.tulane.edu/ -biochem/WW/PepDraw/	Tool to compute net charge and hydrophobicity
Peptide Structure Prediction Server	Pep-Fold	http://bioserv.rpbs.univ-paris-diderot.fr/services/PEP-FOLD/	Tool to predict peptide structures from aminoacid sequence
	PEPstrMod	http://osddlinux.osdd.net/raghava/pepstrmod	Server to predict the tertiary structure of small peptides
Protein Structure Prediction Server Mainly software for the study of protein-ligand interaction	I-TASSER, Discovery Stuio, Atodock vina, Schrödinger Dock, FlexX, ICM-Docking, GOLD	http://zhanglab.ccmb.med.umich.edu/I.TASSER/	Protein structure and function prediction

Source: Tu et al., 2018

For example, BIOPEP (developed at the University of Warmia and Mazury, Poland) not only interlinks databases of protein sequences, bioactive peptides, and sensory peptides, but also involves inbuilt a program that aids in the prediction of proteolytic hydrolysates and allergenic peptides. BIOPEP ("enzyme action" module, http://www.uwm.edu.pl/biochemia/index.php/en/biopep) and ExPASy-PeptideCutter (http://web.expasy.org/peptide_cutter) are popular online tools for predicting amino acids or peptides release from specific protein substrates, based on the knowledge of cleavage specificity of specific enzymes. Furthermore, it is possible to explore several controlled enzymatic hydrolysis methods by using virtual digestion in combination with the bioactive peptides database (such as BIOPEP). This is achieved by: (a) identifying the number of reported peptides with a specific activity among the released peptides *in silico* (Yu et al., 2016; Fu et al., 2016); (b) calculating the frequency of occurrence of food-derived bioactive peptides (BAPs) in a protein chain; (c) predicting the potential biological activity of the protein. *In silico* analysis can not only simulates a single enzyme hydrolysis, but also predicts hydrolysates produced by multiple enzymes. Protein enzymatic hydrolysis under controlled conditions, according to the application of bioinformatics prior to digestion, can greatly improve the experimental efficiency (Tu et al., 2018). Examples of some bioinformatics applications are as follows. Shamim et al. (2018) demonstrated, by *in silico* docking, interactions of a serine metalloprotease (isolated from marine environment of Goa, India) with a HIV-1 gp41 glycoprotein and a cell adhesion protein of *Listeria monocytogenes*. The novel characteristics of this protease allows acting as a potential candidate for various biotechnological and pharmaceutical industries. Ko et al. (2017) isolated angioteinsin I-converting enzyme (ACE) inhibitory peptide from a marine sponge, (*Stylotellaaurantium*) hydrolysate, prepared by various hydrolysis enzymes. The peptides with the highest ACE inhibitory activity were purified and the molecular docking study revealed that ACE inhibitory activity of the purified peptides was mainly attributed to the hydrogen bond interactions and Pi interaction between the dipeptides and ACE. Results suggested that marine sponge *Stylotella aurantium* is a potential raw material for anti-hypertensive nutraceuticals. Roy et al. (2017)

reviewed *in silico* methods for the design of biotherapeutic proteins and peptides, with special attention to antibodies and vaccines. The authors included a comprehensive list of all biotherapeutics approved for clinical use and an overview of prediction methods for binding affinity, cell penetration ability, half-life, solubility, immunogenicity and toxicity of the designed therapeutics.

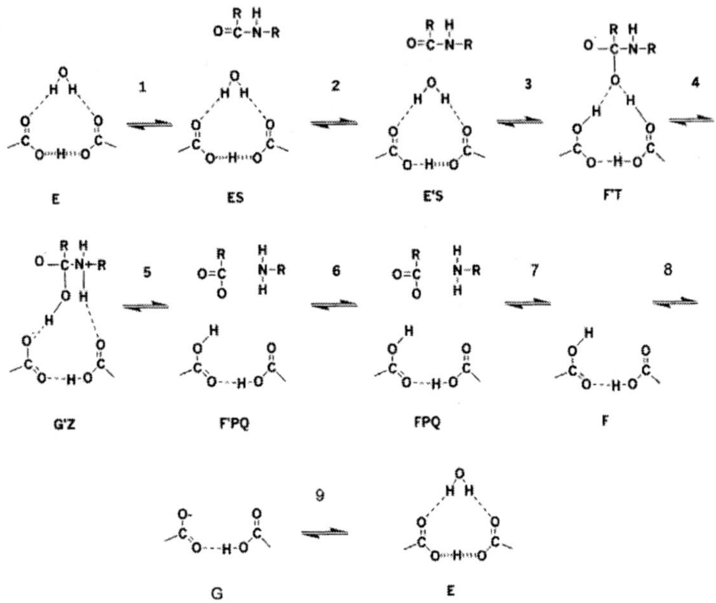

Figure 3. Catalytic mechanism proposed by Northrop (Dunn, 2002).

8. Applications of Marine Proteases and Peptides

Recent biotechnological progresses require the research of new bioactive peptides from natural resources, used as an alternative to chemical products in various applications (food, cosmetics and drugs, among others). Proteases obtained from marine organisms are potential biotechnological materials due to their adaptation to particular, and some time unusual, environments (Kumar and Takagi, 1999; Damare and Raghukumar, 2006). The use of these enzymes results in process efficiency optimization, cost reduction and a wider range of industrial enzyme availability. When specific process requirements involve parameters such as extreme temperatures, saline environments and high pressures, among others, the use of enzymes adapted to such conditions is an obvious advantage. The availability of these enzymes depends on selecting the appropriate source (marine plants, animals or microorganisms) as well as on extraction and purification methods, some of those are still subjects to basic research and biochemical characterization. The search for obtaining proteases from different anatomical regions of marine species has increased during the last years, including those extracted from fish and shellfish viscera and fishing industry wastes.

8.1. Food Industry

The global food industry is one of the largest user of proteases. Some applications of marine enzymes to this industry are protein hydrolysis for source production, pigment extraction, cheese production, flavor enhancement, meat tenderization, fish oil extraction, improved digestibility and/or solubility, detergency, modified functionality such as emulsifying capacity, fat or water binding, gel formation and stability, whipping properties, flavor development, among others (Fernandes, 2014). Several endogenous enzymes were used in model formulations to study disperse food systems, such as protein emulsions and gels (García-Barrientos et al., 2006). These authors found that endogenous proteases of dogfish (*Galeorhinus galeus*) increased viscosity and emulsifying capacity in model systems, although emulsion stability decreased due to unfolding and rearrangement of the proteins forming the emulsion. In the production of food, based on marine proteins, such as surimi gels, proteolytic endogenous enzymes act on myofibrillar proteins, altering the gel structure due the Modori effect, and resulting in a low quality product (Rawdkuen y col., 2004; Stone y Stanley, 1992). In this case, controlling protease activity is necessary.

Debittering and flavor enhancement was achieved by the use of an aminopeptidase from marine *Aspergillus flavus* to cheese processing (Sriram et al., 2012). Heterotopic microorganisms also produced leucyl aminopeptidases that specifically hydrolyze aromatic amino acids. The application of these enzymes is primarily in fish sauce production and milk predigestion for baby food formulations, by detaching peptides on the C-terminal side of lysine and arginine amino acid residues (Steen and Arnosti, 2013).

Meat tenderization is mainly achieved by collagen hydrolysis and myofibrils fragmentation. However, meat protein extensive degradation promotes off-flavors or sticky, lime-like texture. Collagenases acting at low temperature have the advantage of preventing the extensive degradation that occurs with vegetable enzymes, generally used as meat tenderizers, or allowing the growth of undesirable microorganisms due to their peak activity at *circa* 20°C (Fernandes, 2014). *Pseudoalteromonas* sp., a deep-sea psychrophile, produces subtilase, a collagenase active at 0°C, autolyzing 40°C. In addition to tenderizing, this enzyme increases water retention and redness in beef (Chen et al., 2003). Myofibril fragmentation occurs due to endogenous enzyme action during post-rigor conditioning, increasing desirable tenderness and water holding capacity. Texture improvement can be also achieved by an extracellular serine protease also produced by a deep-sea psychrophilic *Pseudoalteromonas* sp. (Zhao et al., 2012).

8.2. Biotechnology Processes

Extensive studies on natural pigment production are published. Coloring materials, such as astaxanthin extracted from crustaceans, are highly demanded by concerned consumers. Enzymatic astaxanthin extraction from shrimp wastes was reported by Armenta and Guerrero Legarreta (2008), and Guerrero Legarreta and Armenta (2010). The authors reported the extraction of carotenoproteins from fermented shrimp waste, by hydrolysis with an enzyme obtained from same shrimp hepatopancreas. The protein-free form of astaxanthin derived the carotenoprotein is of interest for salmon culture, and in natural health products and cosmetics. Furthermore, fermented carotenoproteins could be used in human and animal diets due to their high essential amino acids concentration.

Although around 75% of the fish weight consists of skin, bones, head, and scales, they are often discarded as by-products by the seafood processing industries (skin, bones, and scales). These by-products are a rich source of collagen with a variety of bioactivities. Acid solubilization and pepsin solubilization are major methods for isolating collagen from various parts of fish species (e.g., skin, bones, and scales). For the acid-soluble collagen (ASC) method, 0.5 M acetic acid is used to digest the fish skin in sufficient time, whereas 10% w/v pepsin is used for the pepsin-soluble collagen (PSC) method. Table 1 shows a list of some important marine fish species used for collagen isolation. The PSC method yields higher amounts of collagen as compared to the ASC method (Lugo Magaña et al., 2018). This implies that pepsin in the PSC method is more efficient in digesting skin or bone tissues as compared to acid solution in the ASC method.

Biochemical characteristics of crude alkaline protease extracts from the viscera of goby (*Zosterisessor ophiocephalus*), thornback ray (*Raja clavata*), and scorpionfish (*Scorpaena scrofa*) were studied. At least four caseinolytic proteases bands were observed in zymogram of each enzyme preparation. These proteolytic preparations were successfully used in the deproteinization of shrimp wastes for chitin extraction (Miled et al., 2011). Armenta et al., (2002) also reported the use of proteases extracted from shrimp hepatopancreas in shrimp waste deproteinization in chitin extraction and purification.

A trypsin was purified from the pyloric ceca of Pacific cod (*Gadus macrocephalus*) using chromatographic methods. The cod trypsin was successfully applied to catalyze dipeptide synthesis using a series of "inverse substrates." A mackerel trypsin was also purified from defatted viscera by supercritical carbon dioxide, and its biochemical properties were studied (Miled et al., 2011).

8.3. Pharmaceutical and Cosmeceutical Industry

The need for efficient drugs and an increasing interest in health, well-being, and physical appearance has resulted in high demand for pharmaceuticals and cosmetics (Martins et al., 2014). Cosmeceuticals are cosmetic products with bioactive ingredients, reported to have medical benefits (Venkatesan et al., 2017). They are based on a combination of cosmetics with pharmaceuticals, and biologically active ingredients. A number of bioactive proteins and peptides are obtained from marine organisms, via chemical or enzymatic hydrolysis, for the development of pharmaceutical and cosmeceutical products. The marine environment represents a rich source for enzymes for these industries. Antioxidant, anti-inflammatory, melanin synthesis reduction, tyrosinase inhibition, MMP inhibitor tests are important in the development of cosmeceuticals against aging and skin wrinkling skin (Venkatesan et al., 2017).

Trypsin obtained from cod is the basis of commercial products used to reduce psoriasis, eczema, surgical incisions and skin infections and to prevent cold (Zymetech®, 2019). Likewise, enzymes in fish eggs have been also used as cosmeceuticals. A product for skin exfoliation (Aqua Bio Technology, 2019) contains a serine endo-protease obtained from salmon eggs. The enzyme is recovered from water after the hatching. *In vivo*, these serine endopeptidases is involved in egg hatching by digesting the fibrous protein structure.

Marine fish-derived peptides have been extensively explored for cosmeceutical applications due their properties including antioxidant, antimicrobial, photo-protective, and anti-photo aging activities (Thomas and Kim, 2013). Their biological activities led to the development of several types of anti-aging, skin care, and anti-wrinkle products.

Collagen obtained from marine sources using marine proteases with collagenolytic activity (collagenases), is widely used in the cosmetic industry, as well as in the field of organ transplant and tissue extraction (Prashant and Padma, 2018). Marine fish collagen has skin hydration properties, without odor and with high mechanical strength (Venkatesan et al., 2017). Collagen polypeptides also provided moisture absorption and retention. *In vivo* studies demonstrated that these peptides provided protective effects against UV-induced wrinkle formation and destruction of undesirable skin structures. The mechanism involves increasing the immune reactions, decreasing the loss of moisture and lipid, and repairing endogenous collagen and elastin protein fibers (Hou et al., 2012).

Fish peptides and fish protein hydrolysates have been studied for their photo protective and anti-photo aging activity. Skin, a chemical and physical barrier to protect the body against harmful foreign pollutants, is made up of three different layer: epidermis, dermis and hypodermis (Venkatesan et al., 2017). It is damaged in various ways by some chemicals, ultraviolet (UV) light exposure, and temperature changes; photo aging and inflammation are often caused by UV radiation (Han et al., 2004). Utilization of marine fish-derived peptides for skin protection has been explored. Among them, fish skin and jellyfish (*Rhopilema esculentum*) collagen hydrolysates (Zhuang et al., 2009; Leone et al., 2015). Gelatin hydrolysates from salmon skin protect skin from photoaging (Chen, et al., 2016), and hydrolysate obtained from Pacific cod (*Gadus macrocephalus*) showed protective effects from UV radiation-induced inflammation and collagen reduction in photo aging (Chen and Hou, 2016). Oral administration of gelatin hydrolysate suppressed UV radiation-induced damage to the skin by inhibiting the depletion of endogenous antioxidant enzyme activity (Chen and Hou, 2016). The effects of gelatin hydrolysate extracted from the Pacific cod (*Gadus macrocephalus*) skin was also studied on mouse skin (Chen and Hou, 2016).

Matrix metalloproteinases (MMP) are enzymes that contribute to the extracellular matrix degradation and play major role in normal and pathological tissue remodeling. Imbalance in the expression of MMP leads to severe pathological condition that could initiate cardiac, cartilage, and cancer-related diseases (Thomas and Kim, 2010). Therefore, MMP inhibitors prevents the formation of wrinkles. The marine environment is a rich source of MMP inhibitors. Ngo et al. (2010) reported the isolation of peptides from seahorses that suppresses collagenases. MMP inhibitors were obtained from Atlantic cod muscle (Lødemel et al., 2004) and shark cartilage (González et al., 2001).

Age-related skin thinning is involved in a decrease in the content of collagen in the skin. In addition, wrinkles are a typical symptom of skin aging, and are associated with a decrease in collagen content, reducing skin elasticity of the skin dermal tissues. Since collagen fibers and other extracellular matrix are readily degraded by matrix metalloprotease (MMP), formation of wrinkles is closely associated with increased expression of MMP throughout the skin aging (Thomas and Kim, 2013).

Other applications of marine proteases in medical therapy is the use of proteases of *Octopus vulgaris, Palinurus elephas, Eriphia verrucosa* and *Maja squinado* in cell therapy and pancreas islet extraction (Salamone et al., 2012). Ni et al. (2009) cloned a protease from marine *Aureobasidium pullulans* and expressed in *Yarrowia lipolytica*, resulting in angiotensin-converting enzyme (ACE) inhibitory activity with antihypertensive effects without side effects and antioxidant activity.

8.4. Protein Hydrolysates and Bioactive Peptides

Protein hydrolysates are applied in a number of industrial processes. These are breakdown products of protein enzymatic conversion into smaller peptides, normally containing 2–20 amino acid residues. Although proteins from various sources are used as raw material, fish proteins are a valued resources, due that the fish processing industry produces more than 60% by-products as waste, including skin, heads, viscera, trimmings, and bones (Chalamaiah et al., 2012). These by-products with high protein content and good amino acid balance, are normally processed into low market-value products, such as animal feed, fish meal and fertilizer.

Either obtained by an applied hydrolytic process, or naturally present, have in many marine species, possess a broad spectrum of bioactivities, such as antimicrobial, immunomodulatory, and antioxidant, among others (Cheung et al., 2015).

There are a number of reports in the literature regarding the activity of a wide variety of proteolytic extracts obtained from marine organisms for hydrolysates production. The use of exogenous enzymes, mostly of microbial origin, is preferred due to better process control and higher rate (Sila and Bougatef, 2016). It results in less model systems and scaling-up to industrial application of marine proteases.

Antioxidant peptides obtained by this method and reported by several authors. Hernández-Ortiz (2011) studied an enzyme extract obtain from shrimp wastes to produce antioxidant peptides. The author described a 10 kDa enzyme, and applied it to proteinaceous material to obtain antioxidant compounds. Proteases from scombrid intestines, such as mackerel, were applied to sole (*Limanda aspera*) and pout (*Theragra chalcogramma*) bones. The obtained peptides had biological activity inhibiting linoleic acid autoxidation (Jun et al., 2004; Je et al., 2005). Other antioxidant peptides are produced hydrolyzing *Sardinella aurita* wastes with an enzymatic extract obtained from *Sardina pilchardus* (Bougatef et al., 2009).

The use of proteases to obtain antioxidant biopeptides is also reported. Balti et al. (2011) studied the biochemical properties of protein hydrolysates of various molecular weights, prepared from cuttlefish (*Sepia officinalis*) using an alkalase and *Bacillus licheniformis* NH1 proteases. The authors found that both enzymes increased protein solubility over a wide pH range. The antioxidant activities of cuttlefish protein hydrolysates also increase with increasing degree of hydrolysis, also depending on concentration, retarding lipid peroxidation and β-carotene-linoleic acid oxidation. This research concluded that cuttlefish muscle protein hydrolysates had a high nutritional value and can be used as a supplement to poorly balanced dietary proteins.

García-Moreno et al. (2014) evaluated several fishes of the Mediterranean Sea as raw material to obtain protein hydrolysates with antioxidant activity. Sardine and horse mackerel had the highest DPPH (2,2-diphenyl-1-picrylhydrazyl) scavenging activity (DPPH is a common compound used in antioxidant assay), whereas sardine and small-spotted catshark hydrolysates exhibited the highest ferrous chelating activity. Six antioxidant peptides within myosin and actin structures were also identified in sardine and small-spotted catshark.

Table 3. Some marine peptides, sources and inhibited microorganisms

Antimicrobial peptide	Source	Inhibited microorganisms	Reference
Mytimacin-AF	*Achatina fulica* snail	*S. aureus, K. pneumonia*	Zhong et al., 2013
Aurelin	*Aurelia aurita* jellyfish	*Bacillus megaterium Micrococcus luteus,*	Shenkarev et al., 2012
Piscidins	*Oplegnathus fasciatus* rock bream	*S. iniae, V. harveyi and V. ordalii, and weak antimicrobial activity against E. coli, Vibrio alginolyticus and V. campbellii*	Bae, et al. 2016
Piscidins	*Oreochromis niloticus* Nile tilapia	*E. faecalis, V. vulnificus, V. alginolyticus. A. hydrophila, P. aeruginosa*	Peng et al., 2012 Acosta et al, 2013
Hyastatin	*Scylla paramamosain* crab	*C. glutamicum, M. luteus, P. stutzeri, P. fluorescens, A. hydrophila*	Shan et al., 2016
Sphistin	*Scylla paramamosain* crab	*S. aureus, C. glutamicum, B. subtilis, M. lysodeikticus, M. luteus, S. flexneri, P. stutzeri, P. fluorescens*	Chen et al., 2015
Molluscidin	*Crassostrea giga* oyster	*B. subtilis, M. luteus, S. aureus, E. coli, S. enterica, V. parahaemolyticus*	Seo et al., 2013a Seo et al., 2013b
Centrocins	*Echinus esculentus* sea urchin	*C. glutamicum, S. aureus, E. coli, P. aeruginosa*	Solstad et al., 2016
Acipensins	*Acipenser gueldenstaedtii* Russian sturgeon	*E. coli, L. monocytogenes, S. aureus*	Shamova et al., 2014
SJGAP	*Katsuwonus pelamis* skipjack tuna	*B. subtilis, M. luteus, S. aureus, S. iniae, A. hydrophila, E. coli, V. parahaemolyticus*	Seo et al., 2014
Haliotisins	*Haliotis tuberculate* mollusk	*B. subtilis, E. carotovar*	Zhuang et al., 2015
Chionodracine	*Chionodra cohamatus* crocodile icefish	Antarctic psychrophilic bacteria strains *Psychrobacter sp., B. cereus, E. coli*	Buonocore et al., 2012

Taken from Semreen et al., 2018

Bioactivity, particularly in fish derived peptides, is reported being dependent on structure and amino acid composition and sequences (Khora, 2013). Alemán et al. (2011), who studied antioxidant peptides obtained by hydrolysis of tuna and halibut skins, confirmed this fact and gelatins tunics of jumbo flying squid, using several enzymes produce antioxidant peptides. The authors found that the antioxidant ability was related to aminoacid composition; ABTS (2,2'-azino-bis(3-ethylbenzothiazoline-6-sulphonic acid) radical scavenging was negatively correlated to the total hydrophobic amino acids and imino acids content, and FRAP (ferric reducing antioxidant power) was correlated with Hyl (hydroxylysine) and degree of hydroxylation.

Antimicrobial peptides are in almost all marine organisms (Samreen et al., 2018). A group of compounds known to be powerful agents against a variety of pathogens are classified as non-membrane-permeabilizing peptides (Brogden, 2005). They possess a broad spectrum of antimicrobial activity that can vary according to the nature of the pathogens, attributed to several factors: the diversity structures varying from alpha helix to beta strands (Hancock, 2001) and affinity to the membrane lipid composition of the target organism, which, in turn, depends on the peptide hydrophobicity (Zasloff, 1987). However, the application of antimicrobial marine peptides is limited due to stability at a narrow pH range, reduced activity in the presence of several cations, limited information on the activity under certain environments, and high production costs (Thackray and Moir, 2003). Table 3 lists examples of marine antibacterial peptides, reported in the literature.

Regarding immunomodulation effects of marine biopeptides, Yang et al. (2009) studied the immunomodulatory effects of marine oligopeptide preparation (MOP) from Chum Salmon (*Oncorhynchus keta*) by enzymatic hydrolysis on mice. The authors found that oligopeptide significantly enhance the capacity of lymphocyte proliferation induced by concanavalin A, although no significant weight gain, lymphoid organ indices and phagocytosis capacity were also observed. They concluded that the obtained oligopeptides are possible immune stimulants.

9. Proteases from Marine Microorganisms

As the marine environment includes a wide range of habitats, microorganisms adaptation to this immense variety of conditions generated biocatalysts with outstanding properties, with possible applications in several industries. Total protease sales represent more than 60% of all industrial enzyme sales in the world (Zhang and Kim, 2010), and nearly 65% of industrial proteases are isolated from the marine environment.

They are highly adaptable compounds, being active in the presence of denaturing agents and chemicals, high salt concentrations, temperature, and pressure. This stability makes thermostable proteases important in biotechnological and industrial sectors, affording a high concentration of substrate and less risk of microbial contamination (Barzkar et al., 2018).

The collagenolytic action of a marine bacterium, *Clostridium histolycum* has been studied in the extraction of pancreas islets. This collagenase is more active than its counterpart obtained from land invertebrates is, it is also active at 37°C whereas several toxic metabolites are inhibited (Salamone et al., 2012).

Several marine environments, such as those polluted with fishing effluents, contain microorganism-producing enzymes with industrial potential (Sánchez et al., 2004). Several *Vibrio* species, V*ibrio alginolyticus* among them, produce six proteases (Chen et al., 1999) being the most noticeable collagenases with therapeutic applications against fibroproliferative diseases, such as Dupuytren contraction. The application of this enzyme could be used as a non-chirurgical treatment (Bassetto and Vindigni 2016).

Pseudoalteromonas espejiana and *Pseudoalteromonas haloplanktis* proteasas are produced in active oxygen media (Holmstrom and Kjelleberg, 2006).

Industrial processes improve in efficiency and rate when catalysts are active at extreme conditions, the use of proteases active at extreme temperatures is highly recommended in certain biotechnological processes. Psychrophilic and psychrotolerant marine microorganisms is adapted to such environments conditions that contain adapted proteins and enzymes. Microorganisms adapted to low temperatures is a source of enzymes with high catalytic efficiency; it is found in Polar regions, glacier soil and ice, mountains and deep sea (Georlette et al., 2004). The last, deep sea is also a high-pressure environment. Therefore, microorganisms are not only psychrophilic, but also barophilic or barotolerat (Hamamoto and Russell 1998). Microorganisms isolated from the deep sea are of genus *Pseudomonas*, *Pseudoalteromonas* sp, *Vibrio* sp. and *Bacillus subtilis* BUU1 (Zeng et al., 2003; Chen and Luan, 2003; Arnórsdóttir et al., 2005; Uttatree and Charoenpanich, 2016). Several fungi found in this habitat are also protease producers (Damare et al., 2006). *Bacillus subtilis* BUU1, isolated from marine sediments of the Gulf of Thailand produce a 32 kDa active in a wide pH range (3 a 12) optimal at pH 11, and active from 10 to 80°C. Its stability to surfactants and organic solvents makes this protease an alternative to detergent, wastewater treatment, and leather production (Uttatree and Charoenpanicha, 2016).

At very high temperatures, hyperthermophile microorganisms optimal growth is at 80°C (Stetter, 1996). This environment exists in submarine volcanos, thermal sources and hydrothermal vents. Microorganisms belong to the Domains Archaea and Bacteria, source of thermostable enzymes (Atomi et al., 2011). Michels and Clark (1997) studied a 29 kDa-protease from Archea *Methanococcus hannaschii*, a hyperthermophile microorganism from deep sea and methanogenic environment. It was the first highly stable protease isolated at high temperature and pressure. A thermophile, *Pyrococcus furiosus*, also an Archea, was isolated from geothermic sediments at Vulcan Island, Sicily, Italy. Produces the serineprotease pyrolysin with optimal activity at 115°C and a wide pH range, 6.5 to 10.5. It belongs to the subtilases enzyme family, used for diol production in several industrial processes (Eggen et al., 1990).

CONCLUSION

Total protease sales represent more than 60% of all industrial enzyme sales in the world. Proteases are physiologically necessary for living organisms. They are ubiquitous in a wide diversity of land and marine sources, and a number of industrial applications demonstrate the enormous potential of the marine environment as a source or enzymes, obtained from plants, animals and microorganisms. The use of marine proteases is advantageous over similar biocatalysts from warm-blooded animals, since they are high salt tolerant, hyperthermostable, barophilic, cold adaptable, and highly tolerant to pH. Other unique properties include ease of heat denaturation, high molecular activity at low temperatures and ability to catalyze hydrolysis of native proteins, and ease in large-scale cultivation systems. In addition, marine enzymes usually have a stronger proteolytic activity even at low temperature on several substrates, as compared to the commercial and warm-blooded animal enzymes. In recent years, proteolytic enzymes from marine organisms are used as process aids in a variety of industries, such as food, pharmacy, cosmeceuticals, and wastewater treatment, among others. Even though, the use of marine enzymes can be further expanded at industrial level. Lack of precise information on the mode of action, possible allergic reactions, and side effects prevents wider applications. Basic and applied research is still necessary.

REFERENCES

Acosta, J., Montero, V., Carpio, Y., Velazquez, J., Garay, H. E., Reyes, O., Cabrales, A., Masforrol, Y., Morales, A. & Estrada, M. P. (2013). Cloning and functional characterization of three novel antimicrobial peptides from tilapia (*Oreochromis niloticus*). *Aquaculture*, 372–375, 9-18.

Ahmed, Z., Donkor, O., Street, W. A. & Vasiljevic, T. (2015). Calpains- and cathepsins induced myofibrillar changes in post-mortem fish: Impact on structural softening and release of bioactive peptides. *Trends in Food Science and Technology*, *45*(1), 130-146.

Alemán, A., Giménez, B., Montero, P. & Gómez-Guillén, M. C. (2011). Antioxidant activity of several marine skin gelatins. *LWT - Food Science and Technology*, *44*(2), 407-413.

Aoki, T. (2000). Distribution of cathepsins in red and white muscles among fish species. *Fisheries Science*, 66(4), 776-782.

Aqua Bio Technology ASA, Norway. (2019). https://aquabiotechnology.com/.

Armenta, R., Guerrero Legarreta, I. & Huerta, S. (2002). Astaxanthin extraction from shrimp waste by lactic fermentation and enzymatic hydrolysis of the caroprotein complex. *Journal of Food Science*, *67*(3), 1002-1006.

Armenta, R. E. & Guerrero Legarreta, I. (2008). Amino acid profile and enhancement of the enzymatic hydrolysis of fermented shrimp carotenoproteins. *Food Chemistry*, *112*, 310-315.

Arnórsdóttir, J., Kristjánsson, M. M. & Ficner, R. (2005). Crystal structure of a subtilisin-like serine proteinase from a psychrotrophic *Vibrio* species reveals structural aspects of cold adaptation. *The FEBS Journal*, *272*, 832–845.

Ásgeirsson, B. & Bjarnason, J. B. (1993). Properties of elastase from Atlantic cod a cold-adapted proteinase. *Biochimica et Biophysica Acta - Protein Structure and Molecular Enzymology*, *1164*(1), 91-100.

Atomi, H., Sato., T. & Kanai, T. (2011). Application of hyperthermophiles and their enzymes. *Current Opinion in Biotechnology*, *22*, 618-626.

Auld, D. S. (2013). *Metalloproteases. Encyclopedia of Biological Chemistry*. Boston, MA: Academic Press.

Bae, J. S., Jung, J. M., An, C. M., Kim, J. W., Hwang, S. D., Kwon, M. G., Park, M. A., Kim, M. C. & Park, C. I. (2016). Piscidin: antimicrobial peptide of rock bream, *Oplegnathus fasciatus*. *Fish and Shellfish Immunology*, *51*, 136-142.

Balti, R., Bougatef, A., Ali, N. E., Ktari, N., Jellouli, K., Nedjar-Arroume, N., Dhulster, P. & Nasri, M. (2011). Comparative study on biochemical properties and antioxidative activity of cuttlefish (*Sepia officinalis*) protein hydrolysates produced by alcalase and *Bacillus licheniformis* NH1 proteases. *Journal of Amino Acids*, *2011*, 107179.

Barzkar, N., Homaei, A., Roohullah, H. & Seema, P. (2018). Thermostable marine microbial proteases for industrial applications: scopes and risks. *Extremophiles*, *22*(3), 335-346.

Bassompierre, M., Nielsen, H. H. & Børresen, T. (1993). Purification and characterization of elastase from the pyloric caeca of rainbow trout (*Oncorhynchus mykiss*). *Comparative Biochemistry and Physiology Part B: Comparative Biochemistry*, *106*(2), 331-336.

Bougatef, A., Nedjar-Arroune N., Manni, L., Barkia, A., Guillochon D. & Nasri M. (2009). Purification and identification of novel antioxidant peptides from enzymatic hydrolysates of sardinelle (*Sardinellaaurita*) by-products proteins. *Food Chemistry*, *118*(3), 559-565.

Brier, S., Maria, G., Carginale, V., Capasso, A., Wu, Y., Taylor, R. M. & Engen, J. R. (2007). Purification and characterization of pepsins A1 and A2 from the Antarctic rock cod *Trematomus bernacchii*. *FEBS Journal*, *274*(23), 6152-6166.

Brogden, K. A. (2005). Antimicrobial peptides: pore formers or metabolic inhibitors in bacteria. *Nature Reviews Microbiology*, *3*(3), 238-250.

Buonocore, F., Randelli, E., Casani, D., Picchietti, S., Belardinelli, M. C., Pascale, D. D., Santi, C. D. & Scapigliati, G. (2012). A piscidin-like antimicrobial peptide from the icefish *Chionodraco hamatus* (Perciformes: Channichthyidae): molecular characterization, localization and bactericidal activity. *Fish and Shellfish Immunology*, *33*(5), 1183-1191.

Cárdenas-López, J. L. & Haard, N. F. (2009). Identification of a cysteine proteinase from Jumbo squid (*Dosidicus gigas*) hepatopancreas as cathepsin L. *Food Chemistry*, *112*, 442-447.

Carginale, V., Trinchella, F., Capasso, C., Scudiero, R., Riggio, M. & Parisi, E. (2004). Adaptive evolution and functional divergence of pepsin gene family. *Gene*, *333*, 81-90.

Castillo-Yañez, F. J., Pacheco-Aguilar, R., García-Carreño, F. L. & de Navarrete-Del Toro, M. A. (2004). Characterization of acidic proteolytic enzymes from Monterey sardine (*Sardinops sagax caerulea*) viscera. *Food Chemistry*, *85*(3), 343-350.

Chalamaiah, M., Dinesh kumar, B., Hemalatha, R. & Jyothirmayi, T. (2012). Fish protein hydrolysates: Proximate composition, amino acid composition, antioxidant activities and applications: A review. *Food Chemistry*, *135*(4), 3020-3038.

Chen, T., Fan, D. Q., Zhu, K. X., Shan, Z. G., Chen, F. Y., Hou, L., Cai, L. & Wang, K. J. (2015). Mechanism study on a new antimicrobial peptide sphistin derived from the N-terminus of crab histone H2A identified in haemolymphs of *Scylla paramamosain*. *Fish and Shellfish Immunology*, *47*, 833-846.

Chen, T. & Hou, L. (2016). Protective effect of gelatin polypeptides from Pacific cod (*Gadus macrocephalus*) against UV irradiation-induced damages by inhibiting inflammation and improving transforming growth factor-β/Smad signaling pathway. *Journal of Photochemistry and Photobiology B-Biology*, *162*, 633-640.

Chen, T., Hou, H., Lu, J., Zhang, K. & Li, B. (2016). Protective effect of gelatin and gelatin hydrolysate from salmon skin on UV irradiation-induced photoaging of mice skin. *Journal of Ocean University of China*, *15*, 711-718.

Chen, F.R., Liu. P. C. & Lee, K. K. (1999). Purification and partial characterization of a toxic serine protease produced by pathogenic *Vibrio alginolyticus*. *Microbios*, *98*(390), 95-111.

Chen, X. L., Zhang, Y. Z., Gao, P. J. & Luan, X. W. (2003). Two different proteases produced by a deep-sea psychrotrophic strain *Pseudoaltermonas* sp. SM9913. *Marine Biology*, *143*, 989-993.

Cheung, C. F., Ng, T. & Wong, J. (2015). Marine Peptides: Bioactivities and Applications. *Marine Drugs*, *13*(7), 4006-4043.

Damare, V. S. & Raghukumar, S. (2006). Morphology and physiology of the marine straminipilan fungi, the aplanochytrids isolated from the equatorial Indian Ocean. *Indian Journal of Geo-Marine Sciences*, *35*(4), 326-340.

Damare, V. S., Raghukumar, C., Muraleedharan. U. D. & Raghukumar S. (2006). Deep-sea fungi as a source of alkaline and cold-tolerant proteases. *Enzyme and Microbial Technology*, *39*, 172-181.

De Luca, V., Maria, G., De Mauro, G., Catara, G., Carginale, V., Ruggiero, G., Capasso, A., Parisi, E., Brier, S., Engen, J. R. & Capasso, C. (2009). Aspartic proteases in Antarctic fish. *Marine Genomics*, *2*(1), 1-10.

Dornelles, L. P., Deodato de Souza, M. F. D., da Silva P. M., Figueiredo Procópio, T., Salas Roldan, F. R., de Albuquerque Lima, T., Silva de Oliveira, A. P., Zingali, R. B., Guedes Paiva, P. M., Viana Pontual, E. & Napoleão, T. H. (2018). Purification and characterization of a protease from the visceral mass of *Mytella charruana* and its evaluation to obtain antimicrobial peptides. *Food Chemistry*, *245*, 1169-1175.

Dunn, B. M. (2002). Structure and mechanism of the pepsin-like family of aspartic peptidases. *Chemical Reviews*, *102*, 4431-4458.

Eggen, R., Geerling, A., Watts, J. & De Vos, W. M. (1990). Characterization of pyrolysin, a hyperthermophilic serine protease from the archaebacterium *Pyrococcus furiosus*. *FEMS Microbiology Letters*, *71*, 17-20.

Elleuche, S., Schafers, C., Blank, S., Schroder, C. & Antranikian, G. (2015). Exploration of extremophiles for high temperature biotechnological processes. *Current Opinion in Microbiology*, *25*, 113-119.

Fernandes, P. (2014). Marine enzymes and food industry: insight on existing and potential interactions. *Frontiers in Marine Science*, *1*(46), 1-18.

Fu, Q., Yang, N., Gao, C., Tian, M., Zhou, S., Mu, X., Sun, F. & Li, C. (2018). Characterization, expression signatures and microbial binding analysis of cathepsin A in turbot, *Scophthalmus maximus L.* (SmCTSA). *Fish and Shellfish Immunology*, *81*, 21-28.

Fu, Y., Wu, W., Zhu, M. & Xiao, Z. (2016). *In silico* assessment of the potential of patatin as a precursor of bioactive peptides. *Journal of Food Biochemistry*, *40*, 366e370.

García-Barrientos, R., Pérez Chabela, M. L., Montejano, J. G. & Guerrero Legarreta, I. (2006). Changes in pork and shark (*Rhizopriondon terraenovae*) protein emulsions due to exogenous and endogenous proteolytic activity. *Food Research International*, *39*(9), 1012-1022.

García-Moreno, P. J., Batista, I., Pires, C., Bandarra, N. M., Espejo-Carpio, F. J., Guadix, A. & Guadix, E. M. (2014). Antioxidant activity of protein hydrolysates obtained from discarded Mediterranean fish species. *Food Research International*, *65*(Part C), 469-476.

Ge, L., Xu, Y., Xia, W. & Jiang, Q. (2018). Synergistic action of cathepsin B, L, D and calpain in disassembly and degradation of myofibrillar protein of grass carp. *Food Research International*, *109*, 481-488.

Georlette, D., Blaise, V., Collins, T. & D'Amico, S. (2004). Some like it cold: biocatalysis at low temperatures. *FEMS Microbiology Reviews*, *28*(1), 25-42.

Godahewa, G. I., Perera, N. C. N., Lee, S., Kim, M. J. & Lee, J. (2017). A cysteine protease (cathepsin Z) from disk abalone, *Haliotis discus discus*: Genomic characterization and transcriptional profiling during bacterial infections. *Gene*, *627*, 500-507.

González, R. P., Leyva, A. & Moraes, M. O. (2001). Shark cartilage as source of antiangiogenic compounds: from basic to clinical research. *Biological and Pharmaceutical Bulletin*, *24*(10), 1097-1101.

Guerrero Legarreta, I. & Armenta López, R. (2010). *Proceso enzimático para la obtención de astaxantina y proteína de residuos de camarón fermentado* [*Enzymatic process for obtaining astaxanthin and protein from shrimp wastes*]. Mexican Patent 86555 (November 24th, 2010). Mexico City.

Haard, N. F. & Simpson, B. K. (2000). *Seafood Enzymes: Utilization and Influence on Postharvest Seafood Quality*, 191-207. New York: Marcel Dekker.

Hancock, R. E. (2001). Cationic peptides: effectors in innate immunity and novel antimicrobials. *The Lancet Infectious Disease*, *1*(3), 156-164.

Hedstrom, L. (2002). Serine protease mechanism and specificity. *Chemical Reviews*, *102*(12), 4501-4523.

Hernández-Ortiz J. V., García-Barrientos, R., Hernández-Hernández, E. & Guerrero Legarreta, I. (2011). Evaluación de moléculas de naturaleza proteica obtenidas por hidrólisis endógena de la cabeza de camarón (*Litopeneanus vannamei*) almacenada a -20°C [Analysis of proteins, obtained by endogenous hydrolysis of *Litopeneanus vannamei* heads stored at -20°C]. XXXII Encuentro Nacional y 1er. Congreso Internacional AMIDIQ. May 3-6, Riviera Maya, Quintana Roo, Mexico.

Homaeia, A., Lavajoo, F. & Sariri, R. (2016). Development of marine biotechnology as a resource for novel proteases and their role in modern biotechnology. *International Journal of Biological Macromolecules*, *88*, 542-552.

Ibarra-García, L. E., Tovar-Ramírez, D., Rosas, C., Campa-Córdova, A. I. & Mazón-Suástegui, J. M. (2018). Digestive enzymes of the Californian two-spot octopus, *Octopus bimaculoides* (Pickford and McConnaughey,

1949). *Comparative Biochemistry and Physiology, Part B: Biochemistry and Molecular Biology*, *215*, 10-18.

Jayaraman, G., Srimathi, S. & Bjarnason, J. B. (2006). Conformation and stability of elastase from Atlantic cod, *Gadus morhua*. *Biochimica et Biophysica Acta (BBA)-General Subjects*, *1760*(1), 47-54.

Je, J. Y., Park, P. J. & Kim, S. K. (2005). Antioxidant activity of a peptide isolated from Alaska pollack (*Theragra chalcogramma*) frame protein hydrolysate. *Food Research International*, *38*(1), 45-50.

Jun, S. Y., Park, P. J., Jung, W. K. & Kim, S. K. (2004). Purification and characterization of an antioxidant peptide from enzymatic hydrolysate of yellowfin sole (*Limanda aspera*) frame protein. *European Food Research and Technology*, *219*, 20-26.

Khora, S. S. (2013). Marine fish-derived bioactive peptides and proteins for human therapeutics. *International Journal of Pharmacy and Pharmaceutical Sciences*, *5*(Suppl 3), 31-37.

Klomklao, S., Kishimura, H., Yabe, M. & Benjakul, S. (2007). Purification and characterization of two pepsins from the stomach of pectoral rattail (*Coryphaenoides pectoralis*). *Comparative Biochemistry and Physiology, Part B: Biochemistry and Molecular Biology*, *147*(4), 682-689.

Ko, S. C., Jang, J., Ye, B. R., Kim, M. S., Choi, I. W., Park, W. S., Heo, S. J. & Jung, W. K. (2017). Purification and molecular docking study of angiotensin I-converting enzyme (ACE) inhibitory peptides from hydrolysates of marine sponge *Stylotella aurantium*. *Process Biochemistry*, *54*, 180-187.

Krishnamurthy, A. & Belur, P. D. (2018). A novel fibrinolytic serine metalloprotease from the marine *Serratia marcescens* subsp. *sakuensis*: Purification and characterization. *International Journal of Biological Macromolecules*, *112*, 110-118.

Kumar, C. G. & Takagi, H. (1999). Microbial alkaline proteases: from bioindustrial viewpoint. *Biotechnology Advances*, *17*, 561-94.

Leone, A., Lecci, R. M., Durante, M., Meli, F. & Piraino, S. (2015). The bright side of gelatinous blooms: Nutraceutical value and antioxidant properties of three Mediterranean jellyfish (Scyphozoa). *Marine Drugs*, *13*, 4654-4681.

Lødemel, J. B., Egge-Jacobsen, W. & Olsen, R. L. (2004). Detection of TIMP-2-like protein in Atlantic cod (*Gadus morhua*) muscle using two-dimensional real-time reverse zymography. *Comparative Biochemistry and Physiology Part B: Biochemistry and Molecular Biology*, *139*, 253-259.

Lugo Magaña, O., García Barrientos, R., Guzmán García, X. & Guerrero Legarreta, I. (2018). Purification and preliminary characterization of a blue pigment obtained from "cannon ball" jellyfish *Stomolophus meleagris*. *Russian Journal of Marine Biology*, *44*(6), 447-483.

Márquez-Ríos, E., Cota-Arriola, O., Villalba-Villalba, A. G., Ezquerra-Brauer, J. M., Ocaño-Higuera, V. M., López-Corona, B. E. & Torres-Arreola, W. (2016). Chymotrypsin isolation from jumbo squid (*Dosidicus gigas*) hepatopancreas: Partial characterization and effect on muscle collagen. *Food Science and Biotechnology*, *25* (4), 1011-1016.

Martínez, R., Santos, R., Álvarez, A., Cuzón, G., Arena, L., Mascaró, M., Pascual, C. & Rosas, C. (2011). Partial characterization of hepatopancreatic and extracellular digestive proteases of wild and cultivated *Octopus maya*. *Aquaculture International*, *19*(3), 445-457.

Martínez, R., Santos, R., Mascaró, M., Canseco, L., Caamal-Monsreal, C. & Rosas, C. (2012). Digestive dynamics during chyme formation of *Octopus maya* (Mollusca, Cephalopoda). *Aquaculture Research*, *43*(8), 1119-1126.

Martins, A., Vieira, H., Gaspar, H. & Santos, S. (2014). Marketed marine natural products in the pharmaceutical and cosmeceutical industries: Tips for success. *Marine Drugs*, *12*, 1066-1101.

Michels, P. & Clark D. S. (1997). Pressure-enhanced activity and stability of a hyperthermophilic protease from a deep-sea methanogen. *Applied and Environment Microbiology*, *63*(10), 3985-3991.

Miled, N., Nasri, M., Kishimura, H. & Ben Rebah, F. (2011). Proteins and enzymes from marine resources. *Journal of Amino Acids*, *2011*, 594646.

Muhlia-Almazán, A., Sánchez-Paz, A. & García-Carreño, F. L. (2008). Invertebrate trypsins: a review. *Comparative Biochemistry and Physiology, Part B: Biochemistry and Molecular Biology*, *178*, 655-672.

Ngo, D. H., Qian, Z. J., Ryu, B., Park, J. W. & Kim, S. K. (2010). *In vitro* antioxidant activity of a peptide isolated from Nile tilapia (*Oreochromis niloticus*) scale gelatin in free radical-mediated oxidative systems. *Journal of Functional Foods*, *2*, 107-117.

Ni, X., Yue, L., Chi, Z., Li, J., Wang, X. & Madzak, C. (2009). Alkaline protease gene cloning from the marine yeast *Aureobasidium pullulans* HN2-3 and the protease surface display on *Yarrowia lipolytica* for bioactive peptide production. *Marine Biotechnology*, *11*, 81-89.

Patel, S., Homaei, A., El-Seedi, H. R. & Akhtar, N. (2018). Cathepsins: Proteases that are vital for survival but can also be fatal. *Biomedicine and Pharmacotherapy*, *105*, 526-532.

Peng, K. C., Lee, S. H., Hour, A. L., Pan, C. Y., Lee, L. H. & Chen, J. Y. (2012). Five different piscidins from Nile tilapia, *Oreochromis niloticus*: analysis of their expressions and biological functions. *PLoS ONE*, *7*(11), e50263.

Prashant, K. B. & Padma, B. D. (2018). Collagen and collagenolytic proteases: A review. *Biocatalysis and Agricultural Biotechnology*, *15*, 43-55.

Rawdkuen, S., Benjakul, S., Visessanguan, W. & Lanier, T. C. (2004). Chicken plasma protein: Proteinase inhibitory activity and its effect on surimi gel properties. *Food Research International*, *37*(2), 156-165.

Rodríguez-Siordia, I., Rojo-Arreola, L., Navarrete del Toro, M. A. & García-Carreño, F. (2018). American lobster Cathepsin D, an aspartic peptidase resistant to proteolysis and active in organic solvents, non-ionic detergents and salts. *International Journal of Biological Macromolecules*, *107*, 1501-1509.

Rossano, R., Larocca, M., Lamaina, A., Viggiani, S. & Riccio, P. (2011). The hepatopancreas enzymes of the crustaceans *Munida* and their potential application in cheese biotechnology. *LWT - Food Science and Technology*, *44*(1), 173-180.

Roy, A., Nair, S., Sen, N., Soni, N. & Madhusudhan, M. S. (2017). *In silico* methods for design of biological therapeutics. *Methods*, *131*, 33-65.

Saikhedkar, N., Summanwar, A., Joshi, R. & Giri, A. (2015). Cathepsins of lepidopteran insects: aspects and prospects. *Insect Biochemistry and Molecular Biology*, *64*, 51-59.

Sakharov, I. Y., Dzunkovskaya, A. V., Artyukov, A. A. & Zakharova, N. N. (1993). Purification and some properties of elastase from hepatopancreas of king crab *Paralithodes camtschatica*. *Comparative Biochemistry and Physiology Part B: Comparative Biochemistry*, *106*(3), 681-684.

Salamone, M., Cuttitta A., Seidita G., Mazzola S., Bertuzzi F., Ricordi C. & Gjersi G. (2012). Characterization of collagenolytic/proteolytic marine enzymes. *Chemical Engineering Transactions*, *27*, 1-6.

Sánchez, T., León J., Woolcott, J. & Arauco, K. (2004). Proteasas extracelulares producidas por bacterias marinas aisladas de aguas contaminadas con efluentes pesqueros (Extracellular proteases produced by marine bacteria isolated from seawater contaminated with fishing effluents). *Revista Peruana de Biología*, *11*(2), 179-186.

Semreen, M. H., El-Gamal, M. I., Abdin, S., Alkhazraji, H., Kamal, L., Hammad, S., El-Awady, F., Waleed, D. & Kourbaj, L. (2018). Recent updates of marine antimicrobial peptides. *Saudi Pharmaceutical Journal*, *26*(3), 396-409.

Seo, J. K., Lee, M. J., Go, H. J., Kim, G. D., Jeong, H. D., Nam, B. H. & Park, N. G. (2013a). Purification and antimicrobial function of ubiquitin isolated from the gill of Pacific oyster, *Crassostrea gigas*. *Molecular Immunology*, *53*(1-2), 88-98.

Seo, J. K., Lee, M. J., Go, H. J., Kim, Y. J. & Park, N. G. (2014). Antimicrobial function of the GAPDH-related antimicrobial peptide in the skin of skipjack tuna, *Katsuwonus pelamis*. *Fish and Shellfish Immunology*, *36*(2), 571-581.

Seo, J. K., Lee, M. J., Nam, B. H., Park, T. H. & Park, N. G. (2013b). cgMolluscidin, a novel dibasic residue repeats rich antimicrobial peptide, purified from the gill of the Pacific oyster, *Crassostrea gigas*. *Fish and Shellfish Immunology*, *35*(2), 480-488.

Shahidi, F. & Kamil, Y. V. A. J. (2001). Enzymes from fish and aquatic invertebrates and their application in the food industry. *Trends in Food Science and Technology*, *12*(12), 435-464.

Shamim, K., Sharma, J., Mutnale, M., Dubey, S. K. & Mujawar, S. (2018). Characterization of a metagenomic serine metalloprotease and molecular docking studies. *Process Biochemistry*, *71*, 69-75.

Shamova, O. V., Orlov, D. S., Balandin, S. V., Shramova, E. I., Tsvetkova, E. V., Panteleev, P. V., Leonova, Y. F., Tagaev, A. A., Kokryakov & V. N., Ovchinnikova. T. V. (2014). Acipensins—novel antimicrobial peptides from leukocytes of the Russian sturgeon *Acipenser gueldenstaedtii*. *Acta Natura*e, *6*, 99-109.

Shan, Z., Zhu, K., Peng, H., Chen, B., Liu, J., Chen, F., Ma, X., Wang, S., Qiao, K. & Wang, K. (2016). The new antimicrobial peptide SpHyastatin from the mud crab *Scylla paramamosain* with multiple antimicrobial mechanisms and high effect on bacterial infection. *Frontiers in Microbiology*, *7*, 1140.

Shenkarev, Z. O., Panteleev, P. V., Balandin, S. V., Gizatullina, A. K., Altukhov, D. A., Finkina, E. I., Kokryakov, V. N., Arseniev, A. S. & Ovchinnikova, T. V. (2012). Recombinant expression and solution structure of antimicrobial peptide aurelin from jellyfish *Aurelia aurita*. *Biochemical and Biophysical Research Communications*, *429*(1-2), 63-69.

Siklos, M., BenAissa, M. & Thatcher, G. R. (2015). Cysteine proteases as therapeutic targets: does selectivity matter? A systematic review of calpain and cathepsin inhibitors. *Acta Pharmaceutica Sinica B*, *5*, 506-519.

Sila, A. & Bougatef, A. (2016). Antioxidant peptides from marine by products: isolation, identification and application in foods systems. A review. *Journal of Functional Foods*, *21*, 10-26.

Sivakumar, P., Sampath, P. & Chandrakasan, G. (1999). Collagenolytic metalloprotease (gelatinase) from the hepatopancreas of the marine crab, *Scylla serrata*. *Comparative Biochemistry and Physiology, Part B: Biochemistry and Molecular Biology*, *123*, 273-279.

Solstad, R. G., Li, C., Isaksson, J. Johansen, Svenson, J., Stensvåg, K. & Haug, T. (2016). Novel antimicrobial peptides EeCentrocins 1, 2 and EeStrongylocin 2 from the edible sea urchin *Echinus esculentus* have 6-Br-Trp post-translational modifications. *PLoS ONE*, *11*(3), e0151820.

Sriket, C. (2014). Mini Review-Proteases in fish and shellfish: Role on muscle softening and prevention. *International Food Research Journal*, *21*(1), 433-445.

Sriram, N., Priyadharshini, M. & Sivasakthi, S. (2012). Production and characterization of amino peptidase from marine *Aspergillus flavus*. *International Journal of Microbiological Research*, *3*(3), 221-226.

Steen, A. D. & Arnosti, C. (2013). Extracellular peptidase and carbohydrate hydrolase activities in an Arctic fjord (Smeerenburgfjord, Svalbard). *Aquatic Microbial Ecology*, *69*, 93-99.

Stetter, K. O. (1996). Hyperthermophilic procaryotes. *FEMS Microbiology Reviews*, *18*, 149-158.

Stone, A. P. & Stanley, D. W. (1992). Mechanisms of fish muscle gelation. *Food Research International*, *25*(5), 381-388.

Suriya, J., Bharathiraja, S., Krishnan, M., Manivasagan, P. & Kim, S. K. (2016). Chapter 3: Extremozymes from marine Actinobacteria. *Advances in Food and Nutrition Research*, *79*, 43-66.

Tavano, O. L., Berenguer-Murcia, A., Secundo, F. & Fernández-Lafuente, R. (2018). Biotechnological applications of proteases in food technology. *Comprehensive Reviews in Food Science and Food Safety*, *17*, 412-436.

Thackray, P. D. & Moir, A. (2003). SigM, an extracytoplasmic function sigma factor of *Bacillus subtilis*, is activated in response to cell wall antibiotics, ethanol, heat, acid, and superoxide stress. *International Journal of Bacteriology*, *185*(12), 3491-3498.

Thomas, N. V. & Kim, S. K. (2010). Metalloproteinase inhibitors: status and scope from marine organisms. *Biochemistry Research International*, *2010*, 845975.

Thomas, N. V. & Kim, S. K. (2013). Beneficial effects of marine algal compounds in cosmeceuticals. *Marine Drugs*, *11*, 146-164.

Tomanek, L. (2014). Review-Proteomics to study adaptations in marine organisms to environmental stress. *Journal of Proteomics*, *105*, 92-106.

Tu, M., Cheng, S., Lu, W. & Du, M. (2018). Advancement and prospects of bioinformatics analysis for studying bioactive peptides from food-derived protein: Sequence, structure, and functions. *Trends in Analytical Chemistry*, *105*, 7-17.

Uttatree, S. & Charoenpanich, J. (2016). Isolation and characterization of a broad pH- and temperature-active, solvent and surfactant stable protease from a new strain of *Bacillus subtilis*. *Biocatalysis and Agricultural Biotechnology*, *8*, 32-38.

Venkatesan, J., Anil, S., Kim, S. K. & Shim, M. S. (2017). Marine fish proteins and peptides for cosmeceuticals: A review. *Marine Drugs*, *15*(5), 143-161.

Verma, S., Dixit, R. & Pandey, K. C. (2016). Cysteine proteases: modes of activation and future prospects as pharmacological targets. *Frontiers in Pharmacology*, *7*, 107.

Yang, R., Zhang, Z., Pei, X., Han, X., Wang, J., Wang, L., Long, Z., Shen, X. & Li, Y. (2009). Immunomodulatory effects of marine oligopeptide preparation from Chum Salmon (*Oncorhynchus keta*) in mice. *Food Chemistry*, *113*(2), 464-470.

Yu, F., Young, J. F., Løkke, M. M., Lametsch, R., Aluko, R. E. & Therkildsen, M. (2016). Revalorisation of bovine collagen as a potential precursor of angiotensin I-converting enzyme (ACE) inhibitory peptides based on *in silico* and *in vitro* protein digestions. *Journal of Functional Foods*, *24*, 196-206.

Zasloff, M. (1987). Magainins, a class of antimicrobial peptides from *Xenopus* skin: isolation, characterization of two active forms, and partial cDNA sequence of a precursor. *Proceedings of the National Academy of Science*, *84*(15), 5449-5453.

Zeng, R., Zhang, R., Zhao, J. & Lin, N. (2003). Cold-active serine alkaline protease from the psychrophilic bacterium *Pseudomonas* strain DY-A: enzyme purification and characterization. *Extremophiles*, *7*, 335-337.

Zhang, C. & Kim, S. K. (2010). Research and application of marine microbial enzymes: Status and prospects. *Marine Drugs*, *8*(6), 1920-1934.

Zhao, H. L., Chen, X. L., Xie, B. B., Zhou, M. Y., Gao, X. & Zhang, X. Y. (2012). Elastolytic mechanism of a novel M23 metalloprotease pseudoalterin from deep-sea *Pseudoalteromonas* sp. CF6-2: cleaving not only glycyl bonds in the hydrophobic regions but also peptide bonds in the hydrophilic regions involved in cross-linking. *Journal of Biological Chemistry*, *287*(47), 39710-39720.

Zhong, J., Wang, W., Yang, X., Yan, X. & Liu, R. (2013). A novel cysteine-rich antimicrobial peptide from the mucus of the snail of *Achatina fulica. Peptides*, *39*, 1-5.

Zhou, L., Budge, S. M., Ghaly, A. E., Brooks, M. S. & Dave, D. (2011). Extraction, purification and characterization of fish chymotrypsin: A review. *American Journal of Biochemistry and Biotechnology*, *7*, 104-123.

Zhuang, J., Coates, C. J., Zhu, H., Zhu, P., Wu, Z. & Xie, L. (2015). Identification of candidate antimicrobial peptides derived from abalone hemocyanin. *Developmental and Comparative Immunology*, *49*, 96-102.

Zhuang, Y., Hou, H., Zhao, X., Zhang, Z. & Li, B. (2009). Effects of collagen and collagen hydrolysate from jellyfish (*Rhopilema esculentum*) on mice skin photoaging induced by UV irradiation. *Journal of Food Science*, *74*, H183-H188.

Zymetech®. (2019). *Zymetech, ehf*®. Reykjavík, Iceland. https://zymetech.com/.

In: Proteases: Functions, Mechanisms and Uses ISBN: 978-1-53615-854-0
Editor: Nelson Pérez Guerra

Chapter 4

PROTOZOA SECRETE PROTEASES TO DEGRADE COMPONENTS OF THE HOST INNATE INMUNE SYSTEM

Magda Reyes-López[1], Mireya de la Garza[1], Moisés Martínez-Castillo[2], Mineko Shibayama[2] and Jesús Serrano-Luna[1,*]

[1]Departamento de Biología Celular,
Centro de Investigación y de Estudios Avanzados del IPN,
Ciudad de México, México
[2]Departamento de Infectómica y Patogénesis Molecular,
Centro de Investigación y de Estudios Avanzados del IPN,
Ciudad de México, México

ABSTRACT

Parasitic protozoa are capable of causing both acute and chronic infections in humans; as a whole, they are estimated to affect over 500 000 people worldwide, causing malaria, trypanosomiasis, amoebiasis,

[*] Corresponding Author's E-mail: jjserrano07@yahoo.com.mx. Tel.: (+52 55) 57473800 Ext. 5562

leishmaniasis, giardiasis, and trichomoniasis, among other important diseases. To be able to occupy different niches, parasites have developed different strategies to evade the immune response of the host. Within these strategies is the use of secretion proteinases that are able to degrade different components of the innate immune system, such as mucus, which is found mainly in respiratory, digestive and vaginal epithelium, and other proteins, such as degradation of MUC5A; *Acanthamoeba castellanii* degrades lactoferrin, while *Trichomonas vaginalis* has surface and secreted proteinases that degrade mucus and allow it to adhere to epithelial cells. For researchers and clinicians, it is important to understand how protozoan proteases act against the different molecules of the innate immune system to develop strategies that allow the blockage of these molecules, such as the design of specific antiparasitic drugs that impede colonization and imminent parasitism.

Keywords: native immune system, protozoan parasite proteases, mucus, lactoferrin, Immunoglobulins, complement system

1. Introduction

Parasitic protozoa are unicellular eukaryotic pathogens. They are a major cause of global infection and diseases. As a whole, they are estimated to affect approximately 500 000 people worldwide. These organisms typically produce long-lasting chronic infections such as malaria, amoebiasis, giardiasis, etc. although one of these parasites (*Naegleria fowleri*) causes an acute and fatal infection called primary amebic meningoencephalitis (MAP). A critical step in the host interaction is the evasion of the innate immune system, such as mucus, immunoglobulins, lactoferrin, and the complement system. One of the mechanisms that protozoan organisms have developed to evade the innate immune system is the secretion of surface proteases. These proteases are able to degrade different molecules of the immune system. In this review, we describe the interaction between different protozoa proteases with different host substrates. In general, these interactions allow parasites to colonize and invade different host tissues.

2. MUCUS

Mucus is one of the most ancestral defense mechanisms; it regulates homeostasis and prevents the desiccation of metazoan organisms (Bakshani et al. 2018; Lang et al. 2007). Moreover, through the evolution process, mucus has acquired crucial functions in the defense against of foreign particles and potential pathogens (Bakshani et al. 2018; Sheehan et al. 2006). Crude mucus is a complex fluid that is formed mainly of mucin glycoproteins, which give mucus its viscous properties, and a diverse range of antimicrobial molecules, including many nonspecific molecules and secreted immunoglobulins that protect against specific microbial antigens (Allen and Snary 1972; Reid and Clamp 1978; Sellers et al. 1983; Sellers et al. 1988). The mucus layer also contains lipids, antimicrobial peptides, immunoglobulins, lactoferrin (Lf), lysozyme and protease inhibitors. All of these complementary and structural proteins, along with cilia, participate in the elimination and clearance of pathogens (Bansil and Turner 2018). Mucus and its proteins have a high efficiency for limiting and eliminating pathogens from ocular, nasal, pulmonary, intestinal and cervical epitheliums. Nevertheless, some microorganisms, including protozoa, have developed successful strategies to avoid this immunological barrier (Hicks et al. 2000; Sacks and Sher 2002).

Mucus is mainly composed of approximately 22 mucin genes (Linden et al. 2008; Thornton and JK 2004). There are two types of mucins, secreted (Allen and Snary 1972) and transmembrane types (Baldus et al. 2004; Rose and Voynow 2006). The type of mucin expressed and produced is usually in accordance with the organ requirements; more mucin production can be synthetized and maintained in the cell membrane or secreted as a small soluble mucin that can form a complex gel (Bansil and Turner 2018). Secreted mucin glycoproteins are large, heavily O-glycosylated glycoproteins that are assembled into homo-oligomers that give mucus its viscous properties and are essential for normal development, playing important roles in protein secretion, stability, processing and function (Thornton et al. 2008). There are five oligomerizing secreted proteins (MUC2, MUC5AC, MUC5B, MUC6 and MUC19) and one nonoligomerizing secreted protein (MUC7) (Thornton et al. 2008). MUC2 to MUC6 share a common ancestor, and the genes of these mucins are clustered on chromosome 11p15.5 (De Bolós et al. 1995; Gum et al. 1994). MUC2 is localized in the colon and MUC5AC is in the stomach and airways (Gum et al. 1994; Ho et al. 1995). MUC5B is also present in the airways and gallbladder (Wickström et al. 1998), and MUC6 is present in the stomach (De Bolós et al. 1995). The fifth gel-forming mucin is MUC19, localized on chromosome 12 (Chen et al. 2004). The oligomerized mucins form homo-oligomers by intermolecular disulfide bonds that are formed between the cysteine-rich D-domains found at the amino and carboxyl termini of these mucins. N-glycosylation and C-terminal dimerization occur in the endoplasmic reticulum, followed by O-glycosylation in the Golgi apparatus where N-terminal multimerization of the dimers as mucins are packaged into granules in preparation for secretion (Thornton et al. 2008).

Additionally, there is a large family of transmembrane mucins, of which MUC1 and MUC4 play an important biological role in cell to cell and cell to extracellular matrix interactions and in cell signaling (Jonckheere and Van Seuningen 2010). Another mucin, MUC16, lines the endocervix, endometrium, and fallopian tubes. MUC16 has been shown to have enhanced binding to IgG antibodies (Hladik and McElrath 2008). This interaction between antibodies and MUC16 within glycocalyx covering the luminal surface of the endocervix enhances mucus barrier function against viruses, bacteria and protozoa.

Saliva is a dilute aqueous secretion that contains 99% water, mucin glycoproteins, lipids, and antimicrobial protein molecules (Ramasubbu et al. 1991; Raynal et al. 2002). Saliva lubricates and cleans the oral cavity (Zhang et al. 2016), provides an innate immune defense against pathogens (Situ et al. 2003) and plays an important role in the aggregation and clearance of microorganisms (Wei et al. 2007). Mucus is found throughout the entire gastrointestinal tract but varies in thickness, ranging from 700 μm in the stomach and large intestine to 150-300 μm in the small intestine (Atuma et al. 2001). The barrier consists of two layers: a thinner inner mucus layer, which is sterile and physically difficult to dislodge, and a thicker outer mucus layer, which is more easily dispersed and not sterile (Johansson et al.

2011; Johansson et al. 2008). The inner layer is much thinner in the small intestine than in the large intestine, possibly because of nutrient absorption (Atuma et al. 2001). The extracellular mucus barrier has three major components: secreted mucins, nonspecific antimicrobials and specific immunoglobulins (McGuckin et al. 2011). The mucus layer is produced by specialized secretory cells that are found throughout the entire intestinal tract (McGuckin et al. 2011). In the mouth and esophagus, the mucosa consists of a squamous epithelial layer with underlying glands that secrete saliva that lubricates and protects these tissues (McGuckin et al. 2011). The secreted mucin glycoproteins that form the major macromolecular constituents of mucus are produced by specialized mucus or goblet cells found throughout the gastrointestinal tract (McGuckin et al. 2011). The other major secretory cells within the gastrointestinal tract are Paneth cells, which are localized mainly adjacent to stem cells deep in the crypts of the small intestine (Stappenbeck 2009). These cells are identified by their characteristic intracellular granules containing a variety of antimicrobial molecules that are secreted into the mucus to ensure the sterility of the stem cell niche (Stappenbeck 2009). Mucus is continuously removed from the mucosal surface; for example, in the respiratory tract, mucus is cleared by the action of cilia and by coughing. On the other hand, in the gastrointestinal tract, the outer mucus layer is continuously shifted by peristaltic movement. In the human stomach alone, more than 70 different oligosaccharides are carried by mucins, mostly neutral and highly fucosylated, with less sialylation and no sulfate groups, suggesting great diversity in a single organ (McGuckin et al. 2011).

The production of proteases by pathogenic protozoans plays a pivotal role in their biological processes, including metabolism and life cycle transition. Furthermore, some proteases are involved in the evasion of immune responses and invasion of human tissue (Klemba and Goldberg 2002). Mucinases are enzymes with the ability to affect the rheological properties of mucus and, consequently, their proper function (Martínez Castillo et al. 2017; Wiggins et al. 2001).

2.1. *Entamoeba histolytica*

Entamoeba histolytica (*E. histolytica*) causes amoebiasis that can affect the intestine and produce hepatic, lung and cerebral abscess diseases with high morbidity and mortality in developing countries. *E. histolytica* cysts are ingested through contaminated food and water. Then, they transform into trophozoites in the terminal ileum, invading the colonic mucosa and producing ulcerative lesions and dysentery. In some cases, amoebas can migrate to other organs producing extracellular illness, and it has been reported that this parasite secretes glycosidases that can degrade colonic mucin oligosaccharides (Moncada et al. 2005). Subsequently, the same research group demonstrated that the use of antisense sequences for EhCP5 (*E. histolytica*, cysteine protease 5) abolished the degradation of colonic mucin (MUC2) (Moncada et al. 2006). It is known that mucins do not present carbohydrate branches in the N- and C-terminal domains; thus, these sites are susceptible to proteolytic attack (Bansil and Turner 2018). Furthermore, Lidell and Hansson (2006) produced recombinant domains of MUC2 and demonstrated that *E. histolytica* secreted proteases that have a specific target at the C-terminal domain of MUC2, whereas the N-terminal domain is resistant to proteolysis (Moncada et al. 2006). In conclusion, these assays showed that both glycosidases and proteases can cause alterations in the primary structure of mucin, inducing alterations in the structural conformation of colonic mucus and in this manner, promoting amoebic invasion of colonic tissue.

2.2. *Giardia lamblia*

Giardia lamblia (*G. lamblia*) is another gastrointestinal parasite that is a major contributor to diarrheal diseases in humans. The trophozoite stage causes the disease (Ankarklev et al. 2010). Although this parasite is not invasive, it adheres to the brush border of the small intestine, causing alterations in nutrient absorption. These alterations cause diarrhea and malabsorption syndrome (Eckmann 2003). This protozoan has some glycosidases, including α- and β-N-acetylglucosaminidase and α-galactosidase (Connaris and Greenwell 1997). The carbohydrate residues that degrade these glycosidases are part of the structure of mucins (Rose and Voynow 2006); thus, it is possible that *G. lamblia* employs these enzymes to attach and degrade small intestine mucins.

2.3. *Trichomonas vaginalis*

Trichomonas vaginalis (*T. vaginalis*) is a parasitic protozoan that infects the urogenital tract and causes trichomoniasis, which is a sexually transmitted infection from person to person through sexual contact with high worldwide prevalence; however, it is a poorly studied parasite. Disease can be chronic, and 50% of infected women are asymptomatic carriers. During pregnancy, this illness is associated with preterm delivery, babies with low birth weights and membrane ruptures. Men can be either asymptomatic carriers or can present acute symptomatology (Sehgal et al. 2012). *T. vaginalis* lacks the cyst stage and exists as pyriform trophozoites, sometimes with an amoeboid shape. They present four flagella, one of which is within an undulating membrane. Trophozoites colonize the lumen and mucosal surface of the human urogenital tract, where they can obtain nutrients, multiply and challenge host immune surveillance (Petrin et al. 1998; Sehgal et al. 2012).

This parasite can degrade mucins that can affect the viscosity of vaginal mucus and promote bacterial vaginitis. Moreover, these changes in mucus rheology can increase the risk of HIV and other infections (Lai et al. 2009; Mall et al. 2017). The adhesion of this protozoan to epithelial cells (HeLa) was altered by the presence of cysteine protease inhibitors (TLCK, TPCK, and leupeptin) observed in studies done *in vitro*. In addition, the authors demonstrated that this treatment also reduced the contact-dependent cytotoxicity (Arroyo and Alderete 1989). Later, it was demonstrated that the cysteine proteases produced and secreted by this microorganism have proteolytic activity against bovine mucin. Moreover, the pH and optimal temperature of these proteases were evaluated using acrylamide gel copolymerized with bovine mucin. The results showed that these enzymes were active over a pH range of 4.5–7.0 and were inhibited with cysteine proteinases inhibitors (E-64 and iodoacetamide), cysteine trypsin-like serine (antipain and leupeptin) and trypsin-like serine (TLCK) (Lehker and Sweeney 1999). Moreover, it has been identified in *T. vaginalis* as a sialidase with specific activity for 2–3 linked sialic acids; however, their real impact on the alteration of mucus function has not been evaluated (Wiggins et al. 2001).

2.4. Free-Living Amoebas

2.4.1. Naegleria fowleri

Naegleria fowleri (*N. fowleri*) is the etiological agent of fulminant infection called primary amoebic meningoencephalitis (PAM), affecting the central nervous system through the olfactory epithelium. During the first stage of the infection, a large amount of mucus is secreted, and exacerbated inflammation is observed. Then, extensive lytic and necrotic areas of tissues present (Carter 1970; Jarolimet al. 2000; Jarolimet al. 2002; Rojas-Hernández et al. 2004; Shibayama et al. 2003). *N. fowleri* is able to produce both proteases and glycosidases that can degrade MUC5AC, which is the main type of mucin in the nasal cavity (Cervantes-Sandoval et al. 2008a; Martínez Castillo et al. 2017). Interestingly, the authors demonstrated *in vitro* and *in vivo* models that specific antibodies against 94 kDa glycosidase released by *N. fowleri* can extend the life of Balb/c mice until 10 days postinfection (Martínez Castillo et al. 2017). The authors conclude that the mucinases released by *N. fowleri* play an important function during the invasion process of the central nervous system.

3. Lactoferrin

Lactoferrin (Lf) is an iron-binding cationic glycoprotein with a molecular weight of 78 kDa and 690 amino acid residues and is part of the primary mammalian innate immune defense system. It is formed by chemical, physical and soluble or cellular factors that have evolved in close relation to pathogens to protect the host from infection. This glycoprotein is present in colostrum and milk and mucosal secretions such as intestinal, respiratory and reproductive tracts as well as in tears and saliva (Kanwar et al. 2015; Ochoa and Cleary 2009).

It is a multifunctional protein known for its antibacterial, antifungal, antiviral, antimicrobial, antioxidant, anti-inflammatory, anti-allergic, anticancerous and antiparasitic properties. Lf has been isolated from human and bovine milk comprising 15-20% of total protein (Kanwar et al. 2015; Ochoa and Cleary 2009).

The structure of Lf consists of a single polypeptide chain that is folded into two lobes (N and C), each of which binds either Fe^{2+} or Fe^{3+} ions with one carbonate ion (CO^{3-}). Lactoferrin is an important host defense molecule, and at the physiological level, it has antimicrobial, immunomodulatory and antioxidant activities (Kanwar et al. 2015).

Lactoferrin is partially saturated with iron but can be fully saturated with iron from the external environment. It is an important part of the innate immune system (Wakabayashi et al. 2006) and is continuously synthesized and secreted into exocrine fluids. In some stages of cell differentiation, iron-free Lf is stored in the granules of neutrophils. During an infection or inflammatory condition, Lf levels are increased, which makes Lf a biomarker for inflammation (Actor et al. 2009; Drago Serrano et al. 2018; Kanwar et al. 2015; Kruzel et al. 2017; Ortíz Estrada et al. 2012). This protein takes part in the first-line host defense mechanism against pathogens by sequestration of iron at infection sites. Reducing iron makes it unavailable for microbes, enhancing phagocytosis and cell adhesion and controlling the release of proinflammatory cytokines.

Moreover, Lf can block the interaction between pathogens and host cell receptors used for entry (Kanwar et al. 2015; Ochoa and Cleary 2009).

Proinflammatory cytokines are modulated by Lf, either increasing or decreasing, depending on the threat identified by the immune system. It also activates natural killer (NK) cells, antibody-dependent cell cytotoxicity and macrophage production (Kruzel et al. 2017; Actor et al. 2009; Drago Serrano et al. 2018; Kanwar et al. 2015).

Early host defenses in infection processes increase Lf concentration in serum; the sequestration of free iron by Lf starves microorganisms of this essential element for growth and division. Lactoferrin was tested against bacteria and was shown to be effective against *Streptococcus*, *Salmonella*, *Shigella*, *Staphylococcus* and *Enterobacter* by a specific mechanism attributed to iron depletion. A similar effect was observed in protozoan parasites; while Lf is able to inhibit the growth of this organism in the host, the entry to the host is not affected (Actor et al. 2009; Kanwar et al. 2015).

However, some organisms have adapted to use Lf as a source of iron and present specific lactoferrin-binding proteins on their surface because of the necessity of iron for survival (Ortíz Estrada et al. 2012).

3.1. *Entamoeba histolytica*

When trophozoites colonize the large intestine, they produce high concentrations of infiltrated neutrophils around the amoebas (León Sicairos et al. 2005). One amoebic hololactoferrin-binding polypeptide of 90 kDa was reported in E. histolytica trophozoites (León Sicairos et al. 2005). Lactoferrin binding to a receptor was further detected in acidic vesicles, as was found in *Tritrichomonas foetus* (*T. foetus*). Inside the acidic environment of amoebic vesicles, cysteine proteases and acid phosphatases are concentrated. All of these factors contribute to iron release from Lf and the degradation of the protein itself (León Sicairos et al. 2005; Ortíz Estrada et al. 2012; Temesvari et al. 1999).

Several amoebic proteases of 250, 100, 40 and 22 kDa were induced in the presence of Lf at pH 7. The activity of these proteases was increased at pH 4, and they were inhibited by specific cysteine protease inhibitors (León Sicairos et al. 2005). Dissimilar to *T. foetus* extracellular cysteine protease activity was only found in *E. histolytica* total protein but not in supernatants of the culture media (León Sicairos et al. 2005).

3.2. *Trichomonas vaginalis*

Recently, the importance of iron-binding proteins such as Lf in *T. vaginalis* has been reviewed (Sehgal et al. 2012). Amitochondriate protozoa such as *T. vaginalis*, *G. lamblia*, and *E. histolytica* require a high extracellular iron concentration from 50 to 200 μM (Weinberg 1974). Due to the presence of Lf to chelate ferric iron, the availability of this element to the pathogen is limited. The invading pathogen develops several mechanisms to acquire iron from this protein of the host immune system. The equilibrium between the invading pathogens strategies and the regulation of the host immune response will determine the evolution of the disease (Sehgal et al. 2012; Weinberg 1974).

It has been reported that iron is essential for the virulence of trichomonas. This element upregulates adhesins, and, as a result, cytoadherence is augmented (Garcia et al. 2003; Lehker et al. 1991); cysteine proteases involved in complement resistance are also increased (Alvarez Sánchez et al. 2007). Other cysteine proteases related to virulence have been reported (Crouch and Alderete 2001; Hernández Gutiérrez et al. 2004; León Sicairos et al. 2004), and it is well known that Lf concentrations fluctuate during phases of the menstrual cycle. After menstruation, Lf was observed at the highest concentration, iron concentration was lowest and *T. vaginalis* increased its cytotoxicity due to the increase in the activity of cysteine protease 65 (CP65) (Sehgal et al. 2012). It shows a very close relationship between adhesion, Lf, iron, and protease that cannot be disregarded.

3.3. *Tritrichomonas foetus*

T. foetus is an obligate parasitic protozoon that can colonize both male and female reproductive tracts, causing the sexually transmitted disease trichomoniasis in bovines. This parasite causes inflammation of the female reproductive tract, causing vaginitis, cervicitis, and endometritis that can often result in early embryonic death with important effects on the economy

(Grab et al. 2001; Talbot et al. 1991). *T. foetus* has been used to study host responses *in vivo* to understand the host-parasite relationship and to develop a more effective vaccine.

Trichomonads have a one-stage life cycle and no ability to survive outside the host reproductive tract. Specific lactoferrin-binding sites on the *T. foetus* surface have been reported (Tachezy et al. 1996), and a 155 kDa protein has been reported as a specific lactoferrin-binding protein. Binding of lactoferrin to *T. foetus* is regulated by hydrophobic and charge-related interactions beside lactoferrin receptors (Grab et al. 2001). It is very well adapted to its niche on the mucosal surface, and several proteases have been detected in the extracellular medium, and protease secretion seems to be the major activity during parasite growth. It was reported that Lf can be completely digested in a few hours (Garber and Lemchuk-Favel 1989; Lehker et al. 1990; Lockwood et al. 1984; Lockwood et al. 1987).

These proteases cleave several substrates, such as fibronectin and fibrinogen and are also involved in the rupture of proteins involved in nonspecific host defense mechanisms, such as Lf. This protein is present in bovine cervical mucus (Grab et al. 2001) at higher concentrations of 50 to 600 μg/ml; this concentration can increase during the curse of the illness. Neutrophils are concentrated in inflammatory sites where the uptake and the intracellular elimination of microbes are increased. Therefore, protease activity may interfere with inflammatory defense and sequestration of iron by the host (Talbot et al. 1991). This could be an important mechanism for obtaining iron. Cleavage of these proteins near the trichomonad surface could interfere with specific and nonspecific host defense and generate protein degradation products that further compromise host defense.

T. foetus proteinases responsible for Lf degradation are cysteine proteases because they are inhibited with the specific cysteine protease inhibitor E-64 (Garber and Lemchuk-Favel 1989; Lehker et al. 1990; Lockwood et al. 1984; Lockwood et al. 1987; Talbot et al. 1991). Endocytic vesicles with cysteine proteases that contain acid phosphatase (Affonso et al. 1994) are responsible for iron release and Lf degradation.

3.4. Free-Living Amoebas

3.4.1. Naegleria fowleri

This parasite releases several proteases, including naegleriapores A and B (Cervantes-Sandoval, I et al. 2008a; Cervantes-Sandoval, Isaac et al. 2008b; Eisen and Franson 1987), and it was recently reported that two cysteine proteases (100 and 75 kDa) secreted into the medium were capable of degrading bovine and human Lf. It is interesting to observe that these proteases are active at neutral and acidic pH. In the case of human Lf, the same activities were observed at pH 5, while at neutral pH, only the 100 kDa protease was observed. These proteases could be involved in iron acquisition from *N. fowleri* in the initial stages of the disease to allow the colonization of the host (Martínez Castillo et al. 2015).

3.4.2. Acanthamoeba castellanii

A. castellanii is distributed worldwide. Their life cycle has two stages: trophozoite and cyst. This microorganism can complete its life cycle in the environment without a host, but they can be opportunistic pathogens causing granulomatous amoebic encephalitis in immunocompromised patients or cutaneous lesions, sinusoidal infections or amoebic keratitis in immunocompetent patients (Khan 2006; Marciano Cabral and Cabral 2003). Infection with *A. castellanii* is related to an increasing number of immunocompromised patients and contact lens wearers. Unfortunately, effective chemotherapy does not exist, and amoebic keratitis becomes in chronic infection, sometimes producing vision loss (Ramírez Rico et al. 2015).

The primary step of the infection is adhesion to several molecules, such as the 130 kDa mannose-binding protein (Garate et al. 2004), which then undergoes phagocytosis and toxin production (Omaña Molina et al. 2010; Sissons et al. 2005). The next event in corneal tissue invasion is the production of proteases reported as the main virulence factors, such as serine proteases (Hadas and Mazur 1993; Mitra et al. 1995; Serrano Luna et al. 2006) or metalloproteases (Khan et al. 2000), and finally, other cytotoxic activities occur (Leher et al. 1998). There is little information about proteases against iron-binding proteins. Recently, it was reported that this microorganism presents proteases against Lf which gives the parasite the ability to obtain iron from this kind of protein. This could also support the establishment of the infection by contribution as a virulence factor (Ramírez Rico et al. 2015).

Proteases reported in this parasite are cysteine proteases of 56.6, 100 and 120 kDa, and some are secreted in the culture media (60.2 and 100 kDa). The activity is maintained at acidic, alkaline or neutral pH levels. Several proteases are secreted in the culture media, but only 60.2 kDa protease is specific for Lf (Ramírez Rico et al. 2015).

The study of these free-living amoebas is nascent; much work is needed to understand the presence of these Lf-specific cysteine proteases in parasites whose life cycle can develop normally without a mammal host.

The relationship between hosts and parasites has had a very close evolution. Depending on the environmental conditions, both the host and parasite react. This is clearly observed in trichomonas, where iron, adhesins, Lf, and cysteine proteases are extremely regulated. In this parasite, cysteine proteases are downregulated when the external iron concentration is high. In this environment, the parasite obtains iron from a source other than Lf. Nevertheless, when the iron amount is low and the Lf concentration is high, adhesion protein and cysteine protease production are increased to maintain the parasite in its niche and to obtain iron, maintain its virulence and ensure that the primary line of host defense, Lf is weakened.

4. Immunoglobulins

Vertebrates coexist with an extraordinarily dense and diverse community of approximately 500-1000 bacterial species; therefore, hosts have developed highly sophisticated and regulatory mechanisms to ensure the development and fitness of the immune system to prevented chronic inflammation (Hooper and Gordon 2001). IgA is regarded as the primary mucosal antibody and has the ability, along with IgM, to be actively transported across the epithelial cell barrier (Wright 2011). A basic assumption about the mucosal immune system is that the effector arm acts through IgA (Geuking et al. 2012). IgA and IgM are antibodies actively produced in mucosally associated lymphoid tissue and are actively transported across mucosal cells, achieving relatively high concentrations in mucosal secretions (Wright 2011). One of the important functions of the mucosal immune system is the production of secretory IgA (SIgA) antibodies, which are dimers of IgA molecules joined by a J chain and a secretory component (Brandtzaeg 2013; Mestecky et al. 1971). These antibodies are produced by plasma cells derived from activated B lymphocytes (Lycke and Bemark 2017). The gut is the largest producer of this immunoglobulin in the *lamina propria* of the mucosal membrane, where it is then transported through the epithelial cell barrier (Brandtzaeg 2007). SIgA antibodies are important for protecting against pathogens and in regulating the composition of commensal microbiota (Brandtzaeg 2007; Conner and Blutt 2013; Pabst et al. 2016). The majority of mucosal IgA plasma cells are derived from B-cell activation in mucosal associated lymphoid tissues (MALTs) (Brandtzaeg 2007). The majority of MALT is localized along the gut, hence, termed gut-associated lymphoid tissue (GALT), and is strongly influenced by gut microbiota (Macpherson et al., 2015). The GALT comprises several structures, of which the Peyer's patches (PP) are the most important IgA-inductive sites (Craig and Cebra 2008). These are macroscopically visible clusters of follicles in the small intestine. Much less visible clusters exist in the large intestine, termed colonic patches (cP) (O'Leary and Sweeney 1986).

4.1. *Entamoeba histolytica*

E. histolytica trophozoites are characterized by high production of cysteine proteases. These amoebic proteases are considered to be important virulence factors. These effects have been implicated in gut inflammation. It has been reported that *E. histolytica* trophozoites have a neutral cysteine protease that is able to cleave IgA (Kelsall and Ravdin 1993).

This parasite degrades human secretory IgA1 and IgA2 by cysteine proteases associated with the amoebic surface (Garcia Nieto et al. 2008). It has also been reported that an *E. histolytica* neutral cysteine proteinase of 29 kDa was able to cleave IgG in both serum media and serum-free media (Tran et al. 1998).

4.2. *Trichomonas vaginalis*

T. vaginalis secretes numerous proteinases during cultivation (Lockwood et al. 1987; North et al. 1990; Scott et al. 1995). A 60 kDa cysteine proteinase was purified, and it was able to degrade IgA antibodies. In addition, the enzyme also degraded hemoglobin in a dose-dependent manner. These results suggest that the acidic cysteine proteinase of *T. vaginalis* may play a dual role in parasite survival by conferring escape from host humoral defense by degradation of immunoglobulins and in supplying nutrients to parasites through the degradation of hemoglobin.

5. Complement Systems

The complement system (CS) is a major component of the immune system (Holers and Banda 2018). Some important functions of the CS are the clearance of pathogenic microorganisms through specific recognition, opsonization and lysis (Schifferli et al. 1986). The complement system also participates in the clearance of circulating immune complexes (CICs), apoptotic cells, and dead cells (Mevorach et al. 1998; Wickström et al. 1998). The functions the CS are not only limited to serum or plasma but to each and every tissue or organ of the body, which are direct targets of various complement components (Holers and Banda 2018). Most proteins of the CS are normally present in the circulation as an inactive (zymogen) form to be activated via proteolytic processing (Holers and Banda 2018). There are at least three major pathways: the classical pathway, the alternative pathway and the lectin pathway. All of these pathways are activated by various antibodies, molecules or microorganisms (Holers and Banda 2018).

In the classical pathway, the CS is activated through C1q, which binds to the heavy-chain crystallizable fragment (Fc) domain of immunoglobulin IgM, IgG1, IgG2a and IgG2b, and all of these can form CICs when combined with an antigen and complement. C1q follows activation through C1r and C1s. C1s cleaves and activates C4 into C4a and C4b and simultaneously cleaves C2 into C2a and C2b, leading to the formation of the C4b2a complex, which itself cleaves C3 into C3a and C3b. C3b binds to C4b2a to generate the C5 convertase of the CS. This initiates the formation of C5b-9, known as the membrane attack complex (MAC) (Arumugam et al. 2006; Morgan 1999).

The alternative pathway (AP) of activation, originally called the properdin-dependent pathway, consists of four proteins: factor B (FB), factor D (FD), and properdin (Pf and C3) (Hinz Jr et al. 1956; Holers and Banda 2018). This pathway is activated spontaneously through hydrolysis of C3 generating C3 (H2O), which can associate with FB, resulting in the cleavage of FB into Ba and Bb by FD (Hourcade 2006). Properdin is a positive regulator of AP convertase. C3b binds to Bb and can cleave C3 to generate more C3b and C3a. C3 can also combine with another C3b molecule, forming C5 convertase. This molecule can start the formation of MAC after cleaving components of C5 into C5a and C5b (Hourcade 2006; Pangburn 2000).

The recognition components of the lectin pathway (LP), mannose-binding lectins (MBIs), ficolins (FCNs), and collectins (CLs) bind directly to microbial and other surfaces with exposed carbohydrates and N-acetyl groups and activate complement via MBL-associated serine proteases (MASPs). MASPS are different carbohydrates that can lead to complement activation (Matsushita and Fujita 1995, 2001a).

5.1. *Leishmania* spp.

Leishmania spp. cause disease on the continents of Asia, Africa, Europe, and North and South America. During their life cycle, *Leishmania* spp. shuttle between a flagellated promastigote stage in the sand fly vector and a nonmotile amastigote stage in a mammalian host. During blood meals on infected hosts, the sand fly vectors ingest amastigote-laden macrophages. These amastigotes become procyclic promastigotes in the sand fly gut, multiply by binary fission, and then develop into metacyclic promastigotes, the infective stage for mammals. The sand fly vector, upon taking another blood meal, inoculates the metacyclic promastigotes into mammalian host dermis, where they are phagocytized by macrophages. Amastigotes develop from metacyclic promastigotes and multiply inside parasitophorous vacuoles (PV) of the macrophages. Multiplication of promastigotes eventually leads to rupture of the cells and propagation of the parasites to other macrophages (Garcia 2001).

Trypanosomatids are a very diverse group of protozoa. *Trypanosoma cruzi* is transmitted to humans by the reduviid insect through insect fecal contamination of skin wounds or mucous membranes while taking a blood meal. In the human host, this parasite has two stages: the amastigote and the extracellular trypomastigote, which are disseminated in the blood. The intracellular amastigote multiplies inside virtually any cell until it leads to rupture of infected macrophages and the release of amastigotes, which then infect additional macrophages (Garcia 2001). *Leishmania* spp. has a zinc metalloprotease on their surfaces (Chaudhuri et al. 1989; Etges et al. 1986). This enzyme accounts for approximately 1% of the total protein in the promastigotes of *Leishmania major* and *L. mexicana* (Bouvier et al. 1985). This protease weighs 63,000 Daltons, and is important in different stages of the life cycle. It is known as major surface protease (MSP); this zinc protease has also been termed GP63, surface acid protease, promastigote surface protease, and leishmanolysin. Two important roles of MSP in the pathogenesis of leishmaniasis have been studied (Yao et al. 2003). These include resistance to complement-mediated lysis (CML) and facilitation of phagocytosis of extracellular promastigotes by macrophages (Yao et al. 2003). MSPs of promastigotes bind to C3 with high affinity. Proteolysis of the active component C3b to the inactive C3bi inhibits CM (Brandtzaeg 2007). A mutant MSP$^{(-/-)}$ of *L. major* promastigotes exhibited increased sensitivity to CML (Joshi et al. 1998).

5.2. *Tritrichomonas foetus*

Bovine trichomoniasis is a local infection of the reproductive tract that interacts with mucosal host defense. The parasite is susceptible to elimination by the complement system; however, *T. foetus* secretes an extracellular protease that degrades bovine complement C3. The destruction of complement components by organisms such as *T. foetus* may enhance their ability to colonize the vaginal mucosa; thus, proteinase secretion protects this parasite against attack from the complement system (Kania et al. 2001).

CONCLUSION

Extracellular and surface proteases are involved in host-parasite interactions and are considered virulence factors in many bacteria, fungi, and protozoa. They are very important in pathogenesis mechanisms, such as attachment to host cells, and specific and nonspecific defense against iron-binding proteins, such as Lf. The ability to cleave these proteins is related to mechanisms of colonization and resistance to host defense. Little is known about the mechanisms of action of proteases on the different molecules of the immune system. It is of utmost importance for researchers and clinicians to understand these mechanisms to develop different strategies, such as the use of protease inhibitors or the use of antibodies against the action of these proteases, and in this way, interfere with the invasion and colonization of these protozoan parasites.

AKNOWLEDGMENT

Work supported by CONACyT, México, grant 1674231.

REFERENCES

Actor, J., Hwang, S. & Kruzel, M. (2009). Lactoferrin as a natural immune modulator. *Current Pharmaceutical Design*, *15*(17), 1956-1973.

Affonso, A., Benchimol K. M., Ribeiro, C., Lins, U. & De Souza, W. (1994). Further studies on the endocytic activity of *Tritrich omonas foetus*. *Parasitology Research*, *80*(5), 403-413.

Alvarez Sánchez, M. E., Solano González, E., Yañez Gámez, C. & Arroyo, R. (2007). Negative iron regulation of the CP65 cysteine proteinase cytotoxicity in Trichomonas vaginalis. *Microbes and Infection*, *9*(14), 1597-1605.

Allen, A. & Snary, D. (1972). The structure and function of gastric mucus. *Gut*, *13*(8), 666-672.

Ankarklev, J., Jerlström-Hultqvist, J., Ringqvist, E., Troell, K. & Svärd, S. (2010). Behind the smile: cell biology and disease mechanisms of *Giardia* species. *Nature Reviews Microbiology*, *8*, 413-422.

Arroyo, R. & Alderete, J. (1989). *Trichomonas vaginalis* surface proteinase activity is necessary for parasite adherence to epithelial cells. *Infection and immunity*, *57*(10), 2991-2997.

Arumugam, T., Magnus, T., Woodruff, T., Proctor, L., Shiels, I. & Taylor, S. (2006). Complement mediators in ischemia-reperfusion injury. *Clinica Chimica Acta*, *374*(1), 33-45.

Atuma, C., Strugala, V., Allen, A. & Holm, L. (2001). The adherent gastrointestinal mucus gel layer: thickness and physical state *in vivo*. *American Journal of Physiology-Gastrointestinal and Liver Physiology*, *280*(5), G922-G929.

Bakshani, C., Morales-Garcia, A., Althaus, M., Wilcox, M., Pearson, J., Bythell, J. & Burgess, J. (2018). Evolutionary conservation of the antimicrobial function of mucus: a first defence against infection. *NPJ Biofilms and Microbiomes*, *4*, 14-14.

Baldus, S., Engelmann, K. & Hanisch, F. (2004). MUC1 and the MUCs: A family of human mucins with impact in cancer biology. *Critical Reviews in Clinical Laboratory Sciences*, *41*(2), 189-231.

Bansil, R. & Turner, B. (2018). The biology of mucus: Composition, synthesis and organization. *Advanced Drug Delivery Reviews*, *124*, 3-15.

Bouvier, J., Etges, B. & Bordier, C. (1985). Identification and purification of membrane and soluble forms of the major surface protein of *Leishmania promastigotes*. *Journal of Biological Chemistry*, *260*(29), 15504-15509.

Brandtzaeg, P. (2007). Induction of secretory immunity and memory at mucosal surfaces. *Vaccine*, *25*(30), 5467-5484.

Brandtzaeg, P. (2013). Secretory IgA: Designed for anti-microbial defense. *Frontiers in Immunology*, *4*(222), 1-17.

Carter, R. (1970). Description of a *Naegleria* sp. isolated from two cases of primary amoebic meningo-encephalitis, and of the experimental pathological changes induced by it. *The Journal of Pathology*, *100*(4), 217-244.

Cervantes-Sandoval, I., Serrano-Luna, J., García-Latorre, E., Tsutsumi, V. & Shibayama, M. (2008a). Characterization of brain inflammation during primary amoebic meningoencephalitis. *Parasitology International*, *57*(3), 307-313.

Cervantes-Sandoval, I., Serrano Luna, J., García Latorre, E., Tsutsumi, V. & Shibayama, M. (2008b). Mucins in the host defence against *Naegleria fowleri* and mucinolytic activity as a possible means of evasion. *Microbiology*, *154*(12), 3895-3904.

Connaris, S. & Greenwell, P. (1997). Glycosidases in mucin-dwelling protozoans. *Glycoconjugates Journal*, *14*(7), 879-882.

Conner, M. & Blutt, S. (2013). The gastrointestinal frontier: IgA and viruses. *Frontiers in Immunology*, *4*, 402.

Craig, S. & Cebra, J. (2008). Pillars article: Peyer's patches: An enriched source of precursors for iga-producing immunocytes in the rabbit. *J. Exp. Med.* 1971. 134: 188-200. *The Journal of Immunology*, *180*(3), 1295-1307.

Crouch, M. & Alderete, J. (2001). *Trichomonas vaginalis* has two fibronectin-like iron-regulated genes. *Archives of Medical Research*, *32*(2), 102-107.

Chaudhuri, G., Chaudhuri, M., Pan, A. & Chang, K. (1989). Surface acid proteinase (gp63) of *Leishmania mexicana.* A metalloenzyme capable of protecting liposome-encapsulated proteins from phagolysosomal degradation by macrophages. *Journal of Biological Chemistry*, *264*(13), 7483-7489.

Chen, Y., Zhao, Y., Kalaslavadi, T., Hamati, E., Nehrke, K., Le, A., Ann, D. & Wu, R. (2004). Genome-wide search and identification of a novel gel-forming mucin MUC19/Muc19 in glandular tissues. *American Journal of Respitratory Cell and Molecular Biology*, *30*(2), 155-165.

De Bolós, C., Garrido, M. & Real, F. (1995). MUC6 apomucin shows a distinct normal tissue distribution that correlates with Lewis antigen expression in the human stomach. *Gastroenterology*, *109*(3), 723-734.

Drago Serrano, M. E., Campos Rodriguez, R., Carrero, J. C. & de la Garza, M. (2018). Lactoferrin and Peptide-derivatives: Antimicrobial agents with potential use in nonspecific immunity modulation. *Current Pharmaceutical Design*, *24*(10), 1067-1078.

Eckmann, L. (2003). Mucosal defences against *Giardia. Parasite Immunology*, *25*(5), 259-270.

Eisen, D. & Franson, R. (1987). Acid-active neuraminidases in the growht media from cultures of pathogenic *Naegleria fowleri* and in sonicates of rabbit alveolar macrophages. *Biochimica et Biophysica Acta (BBA) - General Subjects*, *924*(2), 369-372.

Etges, R., Bouvier, J. & Bordier, C. (1986). The major surface protein of *Leishmania promastigotes* is a protease. *Journal of Biological Chemistry*, *261*(20), 9098-9101.

Garate, M., Cao, Z., Bateman, E. & Panjwani, N. (2004). Cloning and characterization of a novel mannose-binding protein of *Acanthamoeba. Journal of Biological Chemistry*, *279*(28), 29849-29856.

Garber, G. & Lemchuk-Favel, L. (1989). Characterization and purification of extracellular proteases of *Trichomonas vaginalis*. *Canadian Journal of Microbiology*, *35*(10), 903-909.

Garcia, A., Chang, T., Benchimol, M., Klumpp, D., Lehker, M. & Alderete, J. (2003). Iron and contact with host cells induce expression of adhesins on surface of *Trichomonas vaginalis*. *Molecular Microbiology*, *47*(5), 1207-1224.

Garcia, L. (2001). *Diagnostic Medical Parasitology*. Washington, DC: ASM Press.

Garcia Nieto, R. M., Rico Mata, R., Arias Negrete, S. & Avila, E. E. (2008). Degradation of human secretory IgA1 and IgA2 by *Entamoeba histolytica* surface-associated proteolytic activity. *Parasitology International*, *57*(4), 417-423.

Geuking, M., McCoy, K. & Macpherson, A. (2012). The function of secretory IgA in the context of the intestinal continuum of adaptive immune responses in host-microbial mutualism. *Seminars in Immunology*, *24*(1), 36-42.

Grab, D., Lonsdale-Eccles, J., Oli, M. & Corbeil, L. (2001). Lactoferrin-binding proteins of *Tritrichomonas foetus*. *Journal of Parasitology*, *87*(5), 1064-1070.

Gum, J., Hicks, J., Toribara, N., Siddiki, B. & Kim, Y. (1994). Molecular cloning of human intestinal mucin (MUC2) cDNA. Identification of the amino terminus and overall sequence similarity to prepro-von Willebrand factor. *Journal of Biological Chemistry*, *269*(4), 2440-2446.

Hadas, E. & Mazur, T. (1993). Proteolytic enzymes of pathogenic and non-pathogenic strains of *Acanthamoeba* spp. *Tropical Medicine and Parasitology*, *44*(3), 197-200.

Hernández Gutiérrez, R., Avila González, L., Ortega López, J., Cruz Talonia, F., Gómez Gutierrez, G. & Arroyo, R. (2004). *Trichomonas vaginalis*: characterization of a 39-kDa cysteine proteinase found in patient vaginal secretions. *Experimental Parasitology*, *107*(3), 125-135.

Hicks, S., Theodoropoulos, G., Carrington, S. & Corfield, A. (2000). The role of mucins in host-parasite interactions. Part I - Protozoan parasites. *Parasitology Today*, *16*(11), 476-481.

Hinz Jr, C., Jordan Jr, W. & Pillemer, L. (1956). The properdin system and immunity. IV. The hemolysis of erythrocytes from patients with paroxysmal nocturnal hemoglobinuria. *The Journal of Clinical Investigation*, *35*(5), 453-457.

Hladik, F. & McElrath, M. (2008). Setting the stage: host invasion by HIV. *Nature Reviews Immunology*, *8*, 447-457.

Ho, S., Roberton, A., Shekels, L., Lyftogt, C., Niehans, G. & Toribara, N. (1995). Expression cloning of gastric mucin complementary DNA and localization of mucin gene expression. *Gastroenterology*, *109*(3), 735-747.

Holers, V. & Banda, N. (2018). Complement in the initiation and evolution of rheumatoid arthritis. *Frontiers in Immunology*, *9*(1057), 1-21.

Hooper, L. & Gordon, J. (2001). Commensal host-bacterial relationships in the gut. *Science*, *292*(5519), 1115-1118.

Hourcade, D. (2006). The role of properdin in the assembly of the alternative pathway C3 convertases of complement. *Journal of Biological Chemistry*, *281*(4), 2128-2132.

Jarolim, K., McCosh, J., Howard, M. & John, D. (2000). A light microscopy study of the migration of *Naegleria fowleri* from the nasal submucosa to the central nervous system during the early stage of primary amebic meningoencephalitis in mice *Journal of Parasitology*, *86*(1), 50-55.

Jarolim, L., McCosh, K. & Howard, J. (2002). The role of blood vessels and lungs in the dissemination of *Naegleria fowleri* following intranasal inoculation in mice. *Folia Parasitologica*, *49*(3), 183-188.

Johansson, M., Larsson, J., Hansson, H. & Gunnar, C. (2011). The two mucus layers of colon are organized by the MUC2 mucin, whereas the outer layer is a legislator of host-microbial interactions. *Proceedings of the National Academy of Sciences*, *108*(Supplement 1), 4659-4665.

Johansson, M., Phillipson, M., Petersson, J., Velcich, A., Holm, L. & Hansson, G. (2008). The inner of the two Muc2 mucin-dependent mucus layers in colon is devoid of bacteria. *Proceedings of the National Academy of Sciences*, *105*(39), 15064-15069.

Jonckheere, N. & Van Seuningen, I. (2010). The membrane-bound mucins: From cell signalling to transcriptional regulation and expression in epithelial cancers. *Biochimie*, *92*(1), 1-11.

Joshi, P., Sacks, D., Modi, G. & McMaster, W. (1998). Targeted gene deletion of *Leishmania major* genes encoding developmental stage-specific leishmanolysin (GP63). *Molecular Microbiology*, *27*(3), 519-530.

Kania, S., Reed, S., Thomford, J., BonDurant, R., Hirata, K., Corbeil, R., North, M. & Corbeil, L. (2001). Degradation of bovine complement C3 by trichomonad extracellular proteinase. *Veterinary Immunology and Immunopathology*, *78*(1), 83-96.

Kanwar, J., Roy, K., Patel, Y., Zhou, S., Singh, M., Singh, D., Nasir, M., Sehgal, R., Sehgal, A., Singh, R., Garg, S. & Kanwar, R. (2015). Multifunctional iron bound lactoferrin and nanomedicinal approaches to enhance its bioactive functions. *Molecules*, *20*(6), 9703-9731.

Kelsall, B. & Ravdin, J. (1993). Degradation of human IgA by *Entamoeba histolytica*. *The Journal of Infectious Diseases*, *168*(5), 1319-1322.

Khan, N. (2006). *Acanthamoeba*: biology and increasing importance in human health. *FEMS Microbiology Reviews*, *30*(4), 564-595.

Khan, N., Jarroll, E., Panjwani, N., Cao, Z. & Paget, T. (2000). Proteases as markers for differentiation of pathogenic and nonpathogenic species of *Acanthamoeba*. *Journal of Clinical Microbiology*, *38*(8), 2858-2861.

Klemba, M. & Goldberg, D. (2002). Biological roles of proteases in parasitic protozoa. *Annual Review of Biochemistry*, *71*(1), 275-305.

Kruzel, M., Zimecki, M. & Actor, J. (2017). Lactoferrin in a context of inflammation-induced pathology. *Frontiers in Immunology*, *8*, 1438-1438.

Lai, S., Wang, Y., Wirtz, D. & Hanes, J. (2009). Micro- and macrorheology of mucus. *Advanced Drug Delivery Reviews*, *61*(2), 86-100.

Lang, T., Hansson, G. & Samuelsson, T. (2007). Gel-forming mucins appeared early in metazoan evolution. *Proceedings of the National Academy of Sciences*, *104*(41), 16209-16214.

Leher, H., Silvany, R., Alizadeh, H., Huang, J. & Niederkorn, J. (1998). Mannose induces the release of cytopathic factors from *Acanthamoeba castellanii*. *Infection and Immunity*, *66*(1), 5-10.

Lehker, M., Arroyo, R. & Alderete, J. (1991). The regulation by iron of the synthesis of adhesins and cytoadherence levels in the protozoan *Trichomonas vaginalis*. *The Journal of Experimental Medicine*, *174*(2), 311-318.

Lehker, M., Chang, T., Dailey, D. & Alderete, J. (1990). Specific erythrocyte binding is an additional nutrient acquisition system for *Trichomonas vaginalis*. *The Journal of Experimental Medicine*, *171*(6), 2165-2170.

Lehker, M. & Sweeney, D. (1999). Trichomonad invasion of the mucous layer requires adhesins, mucinases, and motility. *Sexually Transmitted Infections*, *75*(4), 231-238.

León Sicairos, C. R., León Félix, J. & Arroyo, R. (2004). tvcp12: a novel *Trichomonas vaginalis* cathepsin L-like cysteine proteinase-encoding gene. *Microbiology*, *150*(5), 1131-1138.

León Sicairos, N., Reyes López, M., Canizalez Román, A., Bermúdez Cruz, R. M., Serrano Luna, J., Arroyo, R. & de la Garza, M. (2005). Human hololactoferrin: endocytosis and use as an iron source by the parasite *Entamoeba histolytica*. *Microbiology*, *151*(12), 3859-3871.

Lidell, M. & Hansson, G. (2006). Cleavage in the GDPH sequence of the C-terminal cysteine-rich part of the human MUC5AC mucin. *Biochemical Journal*, *399*(1), 121-129.

Linden, S., Sutton, P., Karlsson, N., Korolik, V. & McGuckin, M. (2008). Mucins in the mucosal barrier to infection. *Mucosal Immunology*, *1*, 183-197.

Lockwood, B., North, M. & Coombs, G. (1984). *Trichomonas vaginalis*, *Tritrichomonas foetus,* and *Trichomitus batrachorum*: comparative proteolytic activity. *Experimental Parasitology*, *58*(3), 245-253.

Lockwood, B., North, M., Scott, K., Bremner, A. & Coombs, G. (1987). The use of a highly sensitive electrophoretic method to compare the proteinases of trichomonads. *Molecular and Biochemical Parasitology*, *24*(1), 89-95.

Lycke, N. & Bemark, M. (2017). The regulation of gut mucosal IgA B-cell responses: recent developments. *Mucosal Immunology*, *10*, 1361-1374.

Mall, A., Habte, H., Mthembu, Y., Peacocke, J. & de Beer, C. (2017). Mucus and Mucins: do they have a role in the inhibition of the human immunodeficiency virus? *Virology Journal*, *14*(1), 192.

Marciano Cabral, F. & Cabral, G. (2003). *Acanthamoeba* spp. as agents of disease in humans. *Clinical Microbiology Reviews*, *16*(2), 273-307.

Martínez Castillo, M., Cárdenas Guerra, R. E., Arroyo, R., Debnath, A., Rodríguez, M. A., Sabanero, M., Flores Sánchez, F., Navarro Garcia, F., Serrano Luna, J. & Shibayama, M. (2017). Nf-GH, a glycosidase secreted by *Naegleria fowleri*, causes mucin degradation: an *in vitro* and *in vivo* study. *Future Microbiology*, *12*(9), 781-799.

Martínez Castillo, M., Ramírez Rico, G., Serrano Luna, J. & Shibayama, M. (2015). Iron-binding protein degradation by cysteine proteases of *Naegleria fowleri*. *BioMed Research International*, *2015*, 416712-416712.

Matsushita, M. & Fujita, T. (1995). Cleavage of the third component of complement (C3) by mannose-binding protein-associated serine protease (MASP) with subsequent complement activation. *Immunobiology*, *194*(4), 443-448.

Matsushita, M. & Fujita, T. (2001a). Ficolins and the lectin complement pathway. *Immunological Reviews*, *180*(1), 78-85.

McGuckin, M., Lindén, S., Sutton, P. & Florin, T. (2011). Mucin dynamics and enteric pathogens. *Nature Reviews Microbiology*, *9*, 265-278.

Mestecky, J., Zikan, J. & Butler, W. (1971). Immunoglobulin M and secretory immunoglobulin A: Presence of a common polypeptide chain different from light chains. *Science*, *171*(3976), 1163-1165.

Mevorach, D., Mascarenhas, J., Gershov, D. & Elkon, K. (1998). Complement-dependent clearance of apoptotic cells by human macrophages. *The Journal of Experimental Medicine*, *188*(12), 2313-2320.

Mitra, M., Alizadeh, H., Gerard, R. & Niederkorn, J. (1995). Characterization of a plasminogen activator produced by *Acanthamoeba castellanii*. *Molecular and Biochemical Parasitology*, *73*(1), 157-164.

Moncada, D., Keller, K., Ankri, S., Mirelman, D. & Chadee, K. (2006). Antisense inhibition of *Entamoeba histolytica* cysteine proteases inhibits colonic mucus degradation. *Gastroenterology*, *130*(3), 721-730.

Moncada, D., Keller, K. & Chadee, K. (2005). *Entamoeba histolytica*-secreted products degrade colonic mucin oligosaccharides. *Infection and Immunity*, *73*(6), 3790-3793.

Morgan, B. (1999). Regulation of the complement membrane attack pathway. *Critical Reviews in Immunology*, *19*(3), 1-26.

North, M., Robertson, C. & Coombs, G. (1990). The specificity of trichomonad cysteine proteinases analysed using fluorogenic substrates and specific inhibitors. *Molecular and Biochemical Parasitology*, *39*(2), 183-193.

O'Leary, A. & Sweeney, E. (1986). Lymphoglandular complexes of the colon: structure and distribution. *Histopathology*, *10*(3), 267-283.

Ochoa, T. & Cleary, T. (2009). Effect of lactoferrin on enteric pathogens. *Biochimie*, *91*(1), 30-34.

Omaña Molina, M., González Robles, A., Salazar Villatoro, L. I., Cristóbal Ramos, A. R., González Lázaro, M., Salinas Moreno, E., Méndez Cruz, R., Sánchez Cornejo, M., De la Torre González, E. & Martínez Palomo, A. (2010). *Acanthamoeba castellanii*: Morphological analysis of the interaction with human cornea. *Experimental Parasitology*, *126*(1), 73-78.

Ortíz Estrada, G., Luna Castro, S., Piñaa Vázquez, C., Samaniego Barrón, L., León Sicairos, N., Serrano Luna, J. & de la Garza, M. (2012). Iron-saturated lactoferrin and pathogenic protozoa: could this protein be an iron source for their parasitic style of life? *Future Microbiology*, *7*(1), 149-164.

Pabst, O., Cerovic, V. & Hornef, M. (2016). Secretory IgA in the coordination of establishment and maintenance of the microbiota. *Trends in Immunology*, *37*(5), 287-296.

Pangburn, M. (2000). Host recognition and target differentiation by factor H, a regulator of the alternative pathway of complement. *Immunopharmacology*, *49*(1), 149-157.

Petrin, D., Delgaty, K., Bhatt, R. & Garber, G. (1998). Clinical and Microbiological Aspects of *Trichomonas vaginalis*. *Clinical Microbiology Reviews*, *11*(2), 300-317.

Ramasubbu, N., Reddy, M., Bergey, E., Haraszthy, G., Soni, S. & Levine, M. (1991). Large-scale purification and characterization of the major phosphoproteins and mucins of human submandibular-sublingual saliva. *Biochemical Journal*, *280*(2), 341-352.

Ramírez Rico, G., Martínez Castillo, M., de la Garza, M., Shibayama, M. & Serrano Luna, J. (2015). *Acanthamoeba castellanii* proteases are capable of degrading iron-binding proteins as a possible mechanism of pathogenicity. *Journal of Eukaryotic Microbiology*, *62*(5), 614-622.

Raynal, B., Hardingham, T., Thornton, D. & Sheehan, J. (2002). Concentrated solutions of salivary MUC5B mucin do not replicate the gel-forming properties of saliva. *Biochemical Journal*, *362*(2), 289-296.

Reid, L. & Clamp, J. (1978). The biochemical and histochemical nomenclature of mucus. *British Medical Bulletin*, *34*(1), 5-8.

Rojas-Hernández, S., Jarillo-Luna, A., Rodríguez-Monroy, M., Moreno-Fierros, L. & Campos-Rodríguez, R. (2004). Immunohistochemical characterization of the initial stages of *Naegleria fowleri* meningoencephalitis in mice. *Parasitology Research*, *94*(1), 31-36.

Rose, M. & Voynow, J. (2006). Respiratory Tract mucin genes and mucin glycoproteins in health and disease. *Physiological Reviews*, *86*(1), 245-278.

Sacks, D. & Sher, A. (2002). Evasion of innate immunity by parasitic protozoa. *Nature Immunology*, *3*, 1041-1047.

Scott, D., North, M. & Coombs, G. (1995). The pathway of secretion of proteinases in *Trichomonas vaginalis*. *International Journal for Parasitology*, *25*(6), 657-666.

Schifferli, J., N, Y. & Peters, D. (1986). The role of complement and its receptor in the elimination of immune complexes. *New England Journal of Medicine*, *315*(8), 488-495.

Sehgal, R., Goyal, K. & Sehgal, A. (2012). Trichomoniasis and lactoferrin: Future prospects. *Infectious Diseases in Obstetrics and Gynecology*, *2012*, 536037.

Sellers, L., Allen, A., Morris, E. & Murphy, S. (1983). Rheological studies on pig gastrointestinal mucous secretions. *Biochemical Society Transactions*, *11*(6), 763-764.

Sellers, L., Allen, A., Morris, E. & Ross-Murphy, S. (1988). Mucus glycoprotein gels. Role of glycoprotein polymeric structure and carbohydrate side-chains in gel-formation. *Carbohydrate Research*, *178*(1), 93-110.

Serrano Luna, J. d. J., Cervantes Sandoval, I., Calderón, J., Navarro García, F., Tsutsumi, V. & Shibayama, M. (2006). Protease activities of *Acanthamoeba polyphaga* and *Acanthamoeba castellanii*. *Canadian Journal of Microbiology*, *52*(1), 16-23.

Sheehan, J., Keimer, M. & Pickles, R. (2006). Innate immunity and mucus structure and function. *Novartis Found Symp*, *279*, 155-166.

Shibayama, M., Serrano-Luna, J., Rojas-Hernández, S., Campos-Rodríguez, R. & Tsutsumi, V. (2003). Interaction of secretory immunoglobulin A antibodies with *Naegleria fowleri* trophozoites and collagen type I. *Canadian Journal of Microbiology*, *49*(3), 164-170.

Sissons, J., Kim, K., Stins, M., Jayasekera, S., Alsam, S. & Khan, N. (2005). *Acanthamoeba castellanii* induces host cell death via a phosphatidylinositol 3-kinase-dependent mechanism. *Infection and immunity*, *73*(5), 2704-2708.

Situ, H., Wei, G., Smith, C., Mashhoon, S. & Bobek, L. (2003). Human salivary MUC7 mucin peptides: effect of size, charge and cysteine residues on antifungal activity. *Biochemical Journal*, *375*(1), 175-182.

Stappenbeck, T. (2009). Paneth cell development, differentiation & function: New molecular cues. *Gastroenterology*, *137*(1), 30-33.

Tachezy, J., Kulda, J., Bahníková, I., Suchan, P., Rázga, J. & Schrével, J. (1996). Iron Acquisition from lactoferrin and transferrin. *Experimental Parasitology*, *83*(2), 216-228.

Talbot, J., Nielsen, K. & Corbeil, L. (1991). Cleavage of proteins of reproductive secretions by extracellular proteinases of *Tritrichomonas foetus*. *Canadian Journal of Microbiology*, *37*(5), 384-390.

Temesvari, L., Harris, E., Stanley, S. & Cardelli, J. (1999). Early and late endosomal compartments of *Entamoeba histolytica* are enriched in cysteine proteases, acid phosphatase and several Ras-related Rab GTPases. *Molecular and Biochemical Parasitology*, *103*(2), 225-241.

Thornton, D. & Sheehan, J. K. (2004). From mucins to mucus: toward a more coherent understanding of this essential barrier. *Proceedings of the American Thoracic Society*, *1*(1), 54-61.

Thornton, D., Rousseau, K. & McGuckin, M. (2008). Structure and function of the polymeric mucins in airways mucus. *Annual Review of Physiology*, *70*(1), 459-486.

Tran, V., Herdman, D., Torian, B. & Reed, S. (1998). The neutral cysteine proteinase of *Entamoeba histolytica* degrades IgG and prevents its binding. *The Journal of Infectious Diseases*, *177*(2), 508-511.

Wakabayashi, H., Takakura, N., Yamauchi, K. & Tamura, Y. (2006). Modulation of immunity-related gene expression in small intestines of mice by oral administration of lactoferrin. *Clinical and Vaccine Immunology*, *13*(2), 239-245.

Wei, G., Campagna, A. & Bobek, L. (2007). Factors affecting antimicrobial activity of MUC7 12-mer, a human salivary mucin-derived peptide. *Annals of Clinical Microbiology and Antimicrobials*, *6*, 14.

Weinberg, E. (1974). Iron and susceptibility to infectious disease. *Science*, *184*(4140), 952-956.

Wickström, C., Davies, J., Eriksen, G., Veerman, E. & Carlstedt, I. (1998). MUC5B is a major gel-forming, oligomeric mucin from human salivary gland, respiratory tract and endocervix: identification of glycoforms and C-terminal cleavage. *Biochemical Journal*, *334*(3), 685-693.

Wiggins, R., Hicks, S., Soothill, P., Millar, M. & Corfield, A. (2001). Mucinases and sialidases: their role in the pathogenesis of sexually transmitted infections in the female genital tract. *Sexually Transmitted Infections*, *77*(6), 402-408.

Wright, P. (2011). Inductive/effector mechanisms for humoral immunity at mucosal sites. *American Journal of Reproductive Immunology*, *65*(3), 248-252.

Yao, C., Donelson, J. & Wilson, M. (2003). The major surface protease (MSP or GP63) of *Leishmania* sp. Biosynthesis, regulation of expression, and function. *Molecular and Biochemical Parasitology*, *132*(1), 1-16.

Zhang, C., Cheng, X., Li, J., Zhang, P., Yi, P., Xu, X. & Zhou, X. (2016). Saliva in the diagnosis of diseases. *International Journal of Oral Science*, *8*, 133-137.

In: Proteases: Functions, Mechanisms and Uses ISBN: 978-1-53615-854-0
Editor: Nelson Pérez Guerra

Chapter 5

Novel Proteases Enzyme for Industrial Utilization

Tahereh Jamshidnejad Tosarmandazni, Mohammad Barshan-Teshnizi and Hasan Jalili [*]
Department of Life Science Engineering, University of Tehran, Tehran, Iran

Abstract

With regards to massive industrialization, the demand for enzymes promotes rapidly. Proteases as unique classes of enzymes have various roles in industries. However, some of them are hindered by low catalytic activity and insufficient stability during the different industrial process. Therefore, researchers try to improve the ability of these biocatalysts through the discovery of novel enzyme with higher functionality or increased properties through new technology including protein engineering, metagenomics, immobilization, etc. This chapter is reviewed the emerging technologies for the development of novel industrial proteases.

[*] Corresponding Author's E-mail: hjalili@ut.ac.ir.

INTRODUCTION

Through the recent decade, demand for different enzymes types growing rapidly in the industry. The main reason for this growth is the powerful catalytic property and the ecö-friendly nature of the enzymes compared to their expensive chemical counterparts. Enzymes do not possess any environmentally hazardous nature; therefore, cleaner solutions were achieved with fewer costs in comparison with the synthesis of bulk chemicals and compounds. These fast biocatalysts are probably to be the bright future of many mass productions and industries (Asgher et al. 2018).

In the near future engineered and recombinant enzymes will be available for most of the industrial and pharmaceutical processes at a reasonable price. A large number of enzymes as safe biocatalyst has been commercially used in several industries already. Among them, the proteases (EC 3:4, 11-19, 20-24) are popular attractive enzymes in industries (Tavano et al. 2018).

Proteases that are also termed as proteolytic enzymes, peptidases, proteases, and proteinases are extracellular enzymes, which can be found in all kinds of living organisms, including humans. These enzymes can be distinguished based on their origin i.e., microbial (bacterial, fungal and viral), plant, and animal enzymes. Although there are some enzymes from natural plants extract, microbial production of the enzymes is preferred commonly, due to their fast growth and simplicity for the generation of new recombinant enzymes with desired properties (Asgher et al. 2018; Contesini et al. 2018).

Proteases hydrolyze large proteins to smaller peptides through hydrolyzing the proteins peptide bonds. Regards to this function, the protease have proven to be indispensable tools for industrial applications, biotechnological, biomedical and pharmaceutical research. In addition, proteases have been utilized in various aspects of Therapeutics (Gale et al. 2018), cosmetics (Ahmed et al. 2018), detergent (Ramkumar et al. 2018), cooking, cheese-making, food (Tavano et al. 2018), pharmaceutical (Contesini et al. 2018), meat tenderization (Tavano et al. 2018), leather (Hammami et al. 2018b), silk preparation (More et al. 2018), production of bioactive peptides (Hammami et al. 2018a), etc. (Xu et al. 2015)

More than 60% of the universal industrial enzymes production is proteolytic ones. However, some proteases are affected by low catalytic activity and/or insufficient stability during the various industrial process under harsh conditions like extreme temperatures and pH, high salinity, non-conventional media, organic solvents, detergents, high substrate/product concentrations, and oxidizing compounds. Therefore, their specificity, stability, and activity in a broad range of physical conditions is a concern, and optimization of strategies to obtain maximum stability and activity of this enzyme in possible lowest cost would benefit the industrial process overall (Marathe et al. 2018). These strategies categorize protein engineering, metagenomics, immobilization, microenvironment manipulation, *De novo* design, surface display, fusion enzymes, etc. and the combination of these.

The enormous progression in these field has been made a reality by the fast development of different omics technologies, system biology, biochemistry, structural biology, and other computational tools (Duraikannu and Chandrasekaran 2018; Xu et al. 2015; Ahmed et al. 2018).

In this chapter advance novel methods for improved protease enzyme regards to specificity, stability and catalytic activity were reviewed.

PROTEIN ENGINEERING AND METAGENOMICS ADVANCES

Computational design, modeling of proteins and high throughput screening methods are the robust tools of protein engineering that help scientists to improve the characteristics of different industrial enzymes such as proteases (Xiao et al. 2015).

The main point of protein engineering besides every enzyme manipulation is the improvement of the biochemical and catalytic properties of the proteases in order to improve performance in production and processes. Promising molecular biology techniques for the improvement of one or more properties of proteases focus on the proficiency of these tools for broad applications of them including analytical, biotechnological, and therapeutics. Protein engineering provides several efficient techniques that include rational design (Asgher et al. 2018), directed evolution, and site-directed mutagenesis (including deletions, insertions, or/and recombination) methods (Xiao et al. 2015).

Fundamental progress has been made in improving protease specificity, activity and stability with protein engineering besides other efforts.

Directed Evolution

This is a process similar to natural evolution but at a much faster speed. It is a robust technique to generate tailor-made enzymes for various biocatalytic applications. Scientists work on directed evolution over past decades more and more. In this process, following the principles of natural evolution, random genomic mutations are performed and then the selection based on the desired function is applied to customize protein properties, enhance catalytic activities, or evolve completely new protein catalysts. The major characteristics of directed are its great speed compared to million years of natural evolution speed. (Zeymer and Hilvert 2018).

Directed evolution encompass these steps:

1. Providing of a mutant library
2. Screening of mutants with desirable function
3. Isolation of improved genes

The main strategy for this direction of enzymes is the creation of random mutations of genes followed by the selection of improved mutants with desirable phenotypes. This process is repetitive rounds to obtain the desirable changes in genetic levels and subsequently phenotypes (Rigoldi et al. 2018).

The successful mutation strategies accumulate in substrate binding pockets and catalytic site most of the times. Currently, several mutation technologies are available to introduce the targeted gene mutation or random mutagenesis. The mutant gene is then transformed with cloning and

expression plasmids, into bacteria to produce a recombinant protein (Rigoldi et al. 2018).

Different mutagenesis techniques are used such as chemical mutagenesis, random mutagenesis, saturation mutagenesis, and DNA shuffling. Among these available strategies for the generation of genetic diversity, most studies in directed evolution are focusing on the use of error-prone PCR for introducing mutations. This method has a relatively easy and inexpensive experimental procedure (Xiao et al. 2015).

In order to alter substrate binding affinity, molecular selectivity or catalytic reaction rate of proteases through protein engineering, scientists use different approaches like rational design which generally target residues at active sites or near that and/or change other important amino acids in enzyme operational domain. Besides, directed evolution has been particularly successful for the generation of proteases with activities towards new substrates, enhancing selectivity, or stability, also through active site mutations as well as mutations distributed throughout the protein molecule (Lancaster et al. 2018).

Reproduction of the useful natural biocatalysts for industrial purpose in laboratory conditions is discovered through metagenomics. Thus, to achieve desirable enzymes, scientists use various enzyme engineering, metagenomics and stabilization techniques for application in several commercial applications (Lancaster et al. 2018).

Metagenomics, on the other hand, is the screening of biomolecules from environmental samples, which consist of two main methods: function-based and sequence-based screening of metagenomic libraries. These two approaches are carried out by the construction of metagenomic libraries where fragmented genomic DNA is cloned in suitable expression vectors that let the expression of the required genes. For the identification of novel enzymes encoded by a single gene or a small sized operon, small sized insert clonal libraries are used, whereas large sized insert libraries are necessary for the segregation of large biosynthetic gene clusters, which encode for complex pathways containing multiple genes. *Escherichia coli* are usually used as a heterologous host for the transformation of metagenomic clones obtained after library construction (Poeta et al. 2018).

PROTEASES SPECIFICITY IMPROVEMENTS

Proteases have diverse and complex roles in a wide array of biological processes with extended substrate specificity, yet native proteases generally lack the substrate specificity; they recognize the carboxylic side of an amino acid that forms a peptide bond, regardless of the protein type (Nguyen et al. 2018).

Different classes of proteases, specifically, are categorized based on their mechanism of action, which is composed of families of proteolytic enzymes: Asp (A), Cys (C), Glu (G), Metallo (M), Asn (N), Mixed (P), Ser (S), Thr (T) and Unknown (U). These can be divided into subfamilies. Because of this specification, these enzymes are highly selective in cleaving only a small set of peptide sequences, while not acting on other sequences in the protein (Waldner et al. 2018).

The ability of custom proteases to precisely hydrolyzing relevant targets advances the potential to use manipulated proteases as delicate therapeutics and research tools.

Proteases are in the use of many types of research, therapeutics, reagents, and diagnostic applications depend on proteases use, such as the synthesis of peptides (Antonyan et al. 2018), production of Klenow fragments (Mótyán et al. 2013), digestion of undesirable proteins during nucleic acid and cell isolation and tissue dissociation, preparation of recombinant antibody fragments production, exploration the relationships of the structure-function, affinity tags removal from fusion proteins in recombinant proteins, sequencing of peptides, proteolytic digestion of proteins in proteomics studies, and enzymatic peptide synthesis (Zhang et al. 2018).

With the aid of high-throughput technology, assays allow the isolation of novel protease variants from large libraries of proteases. As mentioned above, this process is called directed evolution. By directed evolution approach, proteases have been reprogrammed to cleave peptide substrates more specifically and with high activity at a variety of different amino acids in special positions.

Efforts to engineer the specificity of proteases are generally centered on structure-based approaches involving the transplantation of specificity between homologous enzymes.

The only limited breakthrough has accomplished how protease specificity is encoded in its structure, and research successes in this subject are scarce. Scientists are not sure how rationally redesigned proteases selectivity. In this regard, Varadarajan et al. (2008) report that selective preference for the P1 residue of a substrate can be altered with minimal amino acid substitutions within the S1 pocket of OmpT, an outer membrane protease from *E. coli*. The conclusion is refreshingly simple and carries far-reaching consequences for protein engineering of even other enzyme systems (Varadarajan et al. 2008).

For the purification of recombinant proteins expressed in *E. coli*, OmpT is like a solution to troubles because it can cleave substrates at Arg-Arg position, similar to the function of kexin and furin in proprotein processing (Varadarajan et al. 2008).

OmpT is built as a 10-stranded antiparallel β-barrel long enough to cross the outer membrane of *E. coli* and to have its catalytic domain to solvent.

The His212-Asp210 pair of OmpT mirrors the sequence found in the serine protease His-Asp-Ser triad, and the Asp83-Asp85 pair resembles the active site of an aspartic protease (Varadarajan et al. 2008).

The Asp83-His212 function is to activate an H_2O molecule for nucleophilic attack on the Arg-Arg peptide bond. The S1 pocket, which is specific for arginine and lysine residues, is shaped like a deep site containing Glu27 and Asp208 at its bottom and Ser223 in its vicinity. Another position, S1′ site, is less well defined, accentuate the rather broad preference for arginine but also lysine, isoleucine, and histidine at this position. (Varadarajan et al. 2008).

They tried to solve the problem of redesigning the P1 specificity of OmpT using a brute-force method that performs saturation mutagenesis and error-prone PCR to change the character of the S1 pocket. The aim is to alter the P1 specificity of OmpT from arginine to glutamate in which the amino acid charge is reversed completely, tyrosine which is hydrophobic or threonine that is small polar. The wild type version of the enzyme shows no catalytic activity toward three residues. Enormous libraries of mutants were scanned using identifying strategy for desired substrate activity against a reference counter selection substrate carrying wild-type preference.

The whole procedure is rather easy and simple. The group succeeded in all three cases and identified mutants of OmpT that cleave at the Glu-Arg (ER-OmpT), Tyr-Arg (YR-OmpT) and Thr-Arg (TR2-OmpT) positions with values of kcat/Km that are nearly the same as that found for cleavage of Arg-Arg by wild type (2×10^4 M^{-1} s^{-1}). The Glu-Arg preference of ER-OmpT should be mentioned since it has no analog in any native protease. Notably, P1 specificity earning comes with boost selectivity, as each mutant indicates minor or no activity toward the Arg-Arg bond preferred by the natural form of the enzyme. For the first time, the P1 specificity of an enzyme has been altered without loss of catalytic activity or the strong selectivity found in wild type proteases (Varadarajan et al. 2008).

ER-OmpT carries the three substitutions E27L D208R S223G. The negative charges of Glu27 and Asp280 at the bottom of the S1 site are replaced by a positively charged arginine with negligible changes in packing, which possibly caused by engagement of glutamate at the P1 position of the substrate. These substitutions reverse the polarity of the S1 site with a +3 net charge change. YR-OmpT carries the substitutions E27W and D208H in the S1 site, and five other replacements: V29P, I170V, Y172V, Y221A, and L265V. The E27W and D208H substitutions make the S1 pocket more hydrophobic, but the alternations in Tyr221, Leu265, and Val29 also enlarge access to the S1 site without subsiding its gain in hydrophobicity. Optimal engagement of tyrosine at P1 needs further repacking of the S1 site with the substitution of Tyr172 beneath residue 27. A supplementary round of error-prone PCR generated TR2-OmpT, which carries the substitutions E27H, D208L and S223D in the S1 pocket and the six additional substitutions V29S, D214N, P243S, W253G, N270Y, and S276G. In this case, the changes within the S1 cavity and its entrance relocate a negative charge and maintain an overall polarity and packing for favorable interaction with threonine at the P1 site. The recipe for altering the P1 specificity of OmpT is not complicated: alternation the polarity of the S1 pocket and repack it by working directly on the residues that define the S1

pocket and its entrance (Glu27, Asp208, Ser223 and Val29, Tyr221, Leu265 respectively). Within the same structure, trypsin-like proteases prefer arginine and lysine side chains at the P1 site, but chymotrypsins prefer tyrosine and tryptophan residues. At the bottom of the S1 pocket, trypsins carry Asp189 and chymotrypsins carry Ser189. Earlier studies have shown that alternation of P1 specificity between trypsin and chymotrypsin requires both the D189S replacement and extended conversion at residues outside the S1 pocket. However, the same strategy cannot reverse chymotrypsin into trypsin. Recent site-directed mutagenesis studies on OmpT directed by Varadarajan et al. (2008) have shown that the S189D substitution in chymotrypsin is good enough to confer trypsin-like specificity when access to the S1 position.

A general approach for predicting peptidase substrates specificity is *de novo* studies, using protein structure modeling and biophysical evaluation of enzyme-substrate complexes. Yet, current approaches for enzyme specificity prediction depend on machine learning of sequence patterns in known data and therefore are limited.

In this regard, Pethe et al. (2017) describe a general *de novo* approach in order to predict peptidase substrates using protein structure modeling and biophysical evaluation of enzyme-substrate complexes. They construct atomic resolution models of many candidate complexes of substrate-enzyme for each of five common proteases models: Ser-, Cys-, Asp- and metalloproteases.

They develop an optimized discriminatory scoring function by enzyme design modules from Rosetta and Amber-MMPBSA and rank putative substrates based on calculated interaction energy with a modeled near-attack conformation of the enzyme active site.

They indicate that the energetic patterns obtained from their simulations can be used to firmly rank and classify known cleaved and uncleaved peptides and that these structural-energetic patterns have greater discriminatory power in comparison with purely sequence-based statistical inference. Combining sequence and energetic patterns using machine-learning algorithms further improves classification performance, and analysis of structural models provides physical insight into the structural basis for the observed specificities. This group of scientists further tested the probable ability of the model by designing and experimentally characterizing the cleavage of four novel substrate motifs for the Hepatitis C virus NS3/4 protease using an *in vivo* assay (Pethe et al. 2017).

The presented structure-based approach is generalizable to other protease enzymes with known or modeled structures and complements existing experimental methods for specificity determination.

As an example of rational design, recently Xiaojing J. Gao (2018) and her group designed a synthetic protein-level circuit which is a powerful new cellular protein-protein regulation system. In their study, it has been shown that manipulated proteases can act as compostable protein components that can together implement various circuit-level functions in mammalian cells (Pethe et al. 2017).

In their scheme, termed CHOMP (circuits of hacked orthogonal modular proteases), input proteases dock with and cleave target proteases to inhibit their function. These components can be connected to develop regulatory cascades, binary logic gates, and dynamic analog signal-processing functions. To illustrate the efficacy of this system, they rationally designed a circuit that induces cell death in response to upstream activators of the Ras oncogene (Pethe et al. 2017).

Michael S. Packer et al. (2017) conducted the phage-assisted continuous evolution (PACE) for TEV protease, which canonically hydrolyzed ENLYFQS, to cleave a very different target sequence, HPLVGHM, that is present in human IL-23. A protease obtained from nearly 2500 generations of PACE consist of 20 mutations, hydrolyzes human IL-23 at the target position, and when mixed with IL-23 in cultures of murine splenocytes inhibits IL-23-mediated immune signaling. They characterize the specificity of this elaborated enzyme, indicating shifted and broadened specificity alternation at the six positions with the target amino acid sequence. Mutational dissection and additional protease specificity studies show the molecular basis of some of these changes. The capability of improving the substrate specificity of a protease at many positions in a short time and provides a basis for the obtaining of custom proteases that

catalytically function on target proteins for biotechnological and therapeutic applications be established in such works (Packer et al. 2017).

Stability Improvement of Proteases

Enhancing the performance and temperature resistance of protease variants is one of the broad aspects of developing improved proteases for industrial utilization.

The properties of different enzymes have become highly optimized for their specific condition through natural selection over the span of hundreds of millions of years, so it is not surprising that most of the enzymes do not perform as well in industrial condition. To overcome this problem, combinatorial strategies like directed evolution, protein engineering, rational design, and immobilization strategies have been very successful in producing enzymes with enhanced stabilities and other properties that make them useful in large-scale applications. One of the most important aspects of proteases in the industry is stability through harsh conditions. Present inventions in the era of stability are mostly related to the application of the enzyme in washing and cleaning agents. Unlike the therapeutic, specificity is not a concern here.

It has now surprisingly been found that a combination of several modifications in alkaline proteases are suitable for detergent or cleaning agents and particularly helpful to improve cleaning performance and/or stability. Among these modifications, protein engineering is used with different subjects and strategies like amino acid substitution, site-directed mutagenesis, C-terminus fusion, N-terminus peptide engineering, etc.

The ability to stabilize different enzymes and proteins offers improvement in a variety of applications. Most efforts in order to enhance the enzyme stability in harsh conditions need various rounds of high-throughput screening of libraries and provide only minimal improvements in its stability. For circumvent the limitation, computational library design is a solution. Hein J. Wijma et al. design a protocol that can increase enzyme stability by 20 – 35°C with fewer than 200 experimental variants. They termed the protocol FRESCO (Wijma et al. 2018).

In FRESCO, first, they scan the entire protein structure to identify stabilizing disulfide bonds and point mutations, and then explore their effect by molecular dynamics simulations, and provide mutant libraries with potent variants to exhibit enhanced stability. After experimental verification, the most effective mutations are combined to produce highly robust enzymes. The invented protease allows more efficient stain removal options from hard surfaces, fabrics, etc. Furthermore, it can tolerate a wide range of temperatures in order to improved cleaning performance, e.g., between 15°C to 90°C (Wijma et al. 2018).

As another example in the era of stability, it is known that removal of Small ubiquitin-like modifier (SUMO) is a common use of proteases. SUMO is a protein fusion domain to facilitate expression and purification of recombinant proteins, and a SUMO-specific protease is needed to remove SUMO afterward. Although this protease is highly specific, the limitation in solubility and stability hamper its utilization as an *in vitro* reagent. Here, is a report of improved SUMO protease enzymes obtained *via* two approaches. First, they developed a computational method and used it to re-engineer native Ulp1 from *Saccharomyces cerevisiae* to improve protein solubility. Second, they discovered an improved SUMO protease via genomic mining of the thermophilic fungus *Chaetomium thermophilum* (Lau et al. 2018).

As the stability and reusability of soluble proteases are of major concerns, which limit their industrial applications, Abdelnasser Salah Shebl Ibrahim et al. (2016) immobilized alkaline protease from *Bacillus* sp. NPST-AK15 onto hollow core-mesoporous shell silica (HCMSS) nanospheres. In their research, Non-, ethane- and amino-functionalized HCMSS nanospheres were synthesized and characterized. NPST-AK15 was immobilized onto the synthesized supports by physical and covalent immobilization methods. Protease immobilization by covalent attachment onto the activated HCMSS–NH2 nanospheres showed 75.6% immobilization yield and loading capacity (88.1 g protein/mg carrier) (Ibrahim et al. 2016).

In comparison to free enzyme, the covalently immobilized protease gained a slight shift in the optimal pH from 10.5 to 11.0, respectively. The optimum temperature for the catalytic activity of the free and immobilized enzyme was 60°C. Notably, when the free enzyme was inactivated completely at 60°C after 1 h, the immobilized enzyme still keeps 63.6% of its initial activity. Besides, the immobilized protease showed 1.6-fold higher V_{max} and k_{cat}, 2.4-fold higher k_{cat}/K_m. In addition, the immobilized protease showed nearly 1.5-fold higher affinity to the substrate. In addition, the enzyme stability in different organic solvents was highly enhanced upon immobilization. The immobilized enzyme showed increased stability in some commercial detergents including OMO, Tide, Ariel, Bonux, and Xra by up to 5.2-fold and the catalytic efficiency of the immobilized protease was maintained for at least twelve consecutive reaction cycles (Ibrahim et al. 2016).

Improvement of Catalytic Activity

Beside the improvement of the thermophilicity and thermostability performance of the protease variants, catalytic activity is usually evaluated through protein engineering and vice versa.

For example, a well-known protease used in degrading insoluble keratin substrates is keratinase which have many advantages to hydrolyze highly hydrophobic substrate compared with the common proteases like papain, trypsin, and pepsin. Nowadays, keratinolytic proteases with more excellent thermostability and catalytic activity are demanded in the industrial application and commercial market.

Zhen Fang et al. (2016) reported a novel keratinase named KerSMD from *S. maltophilia* BBE11-1 with attractive features in textile and waste treatment. Also, based on the model structure and rational design, they did successful alternation of keratinase KerSMD to produce excellent variants of FDD and DDF with obviously improved keratinolytic activity and thermostability (Fang et al. 2016).

Mesophilic alkaline serine proteases have been applied in different industries, but they have an optimal catalytic temperature of around 50°C and their activity decreases sharply at low temperature. To improve their cold activity, it is vital to increase the catalytic performance of alkaline serine proteases at low temperature. In a combined study of directed evolution and site-directed mutagenesis, Zhao and Feng (2018) attempted to improve the cold activity of *Bacillus pumilus* alkaline serine protease (DHAP).

In this study, seven variants (P9S, A1G/K27Q, A38V, A116T, T162I, S182R, and T243S) of DHAP were generated via directed evolution. The results showed that all of the seven variants had increased proteolytic activity at 15°C towards two substrates: casein and synthetic peptide. In comparison with the wild-type enzyme, the thermostability of these variants did not decrease except for T243S. Further analysis indicated that the improvement in catalytic efficiency was largely attributed to an increase in turnover number (k_{cat}). Moreover, the combined variants obtained by site-directed mutagenesis illustrated a further improvement in specific caseinolytic activity and the k_{cat} value for hydrolysis of the synthetic peptide. The combined variants of P9S/K27Q and P9S/T162I gained a nearly 5-fold increase in caseinolytic activity at 15°C and no decrease of thermostability. These data were analyzed and confirmed by structural modeling at last (Zhao and Feng 2018).

ENZYME IMMOBILIZATION

Biocatalysts have several advantages over industrial chemical catalysts, like high selectivity, high turnover, and efficiency, but some of the catalytic processes are limited by the presence of various organic solvents, extreme temperature, lack proper storage conditions. Therefore, the industrial applications of enzymes depend on the introduction of sufficient and stable immobilization technologies.

The enzyme immobilization dramatically boosts the stability, and also reusability of the enzymes and diminishes the cost of the industrial processes while keeping them environmentally friendly.

Immobilized enzymes are able to catalyze the reactions in vastly different conditions. Several inert materials are used in order to provide divers solid supports. However, in the process of adsorption, the active site of enzymes may be blocked by immobilization supports and this reduces the activity of the immobilized enzyme. Adsorption of enzymes on nanoparticles and adsorption entrapment technique have been found to be more effective as compared to surface adsorption.

The enzyme can be covalently bonded to a matrix directly or using a crosslinking. This approach is the most effective method and increases the efficiency and stability of immobilized enzymes.

Nanosupport bound proteases indicated high efficiency of binding and yield of immobilization. Enzymes attached to such support were found highly more resistant to the denaturation caused by a change in pH, temperature, and other harsh condition of the industry. Compared to their free forms, immobilized enzymes were less inhibited by their inhibitors. Furthermore, it has been shown that the proteases attachment to the nanosize supports prevented them from the steric hindrances during substrate binding to the active sites of the enzymes like common proteases immobilized on the bulk supports. Nanomaterial bound proteases have successfully been used for the variety of applications such as analysis of ochratoxins, heavy metals and other bivalents ions, pharmaceutical compounds and peptides, including monoclonal antibodies and the bound enzymes have also been utilized for screening the drugs on the basis of enzyme inhibition assay. Moreover, nanomaterials immobilized protease have shown their high catalytic efficiency in organic solvents.

Zinc oxide nanoparticles are one of the absorption examples for proteases. In a study conducted by Homaei et al. (2018) zinc oxide nanoparticles were synthesized by the chemical precipitation method, and shrimp (*Penaeus vanamei)* protease was immobilized on the zinc oxide. The immobilization of protease on ZnO nanoparticles improved the thermal and long-term stability, also it improved the optimum functional temperature of the enzyme. The optimum pH of the immobilized enzyme was shifted from 7.0 to 8.0 upon immobilization. Furthermore, due to the immobilization, an increased K_M was recorded, whereas its catalytic efficiency was dropped a little as compared to that of free enzyme. These results indicate that the immobilization of shrimp protease on ZnO nanoparticles enhanced its performance in various biotechnological and industrial applications (Diyanat et al. 2018).

In enzyme entrapment, the protease enzyme is attached to support through forces such as ionic interactions, Vander der Waals forces, and hydrogen bonds. Entrapment is a preferable method as is easy, cheap and produces are often stable. Both hydrophobic and hydrophilic materials are used for enzyme entrapment. Hydrophobic materials are better as they can entrap more enzyme, a higher degree of immobilization and much activity is retained. The enzyme is usually trapped in insoluble beads (calcium alginate) or hydrophobic sol-gels (propyletrimethoxysilane (PTMS) and tetramethoxysilane (TMOS)).

Shah Ali UlQader et al. (2018) showed that the catalytic performance of an immobilized protease enhanced by using entrapment technique. In their study agar-agar, is subjected to entrap serine-protease produced by *Aspergillus niger* KIBGE-IB36. The results showed that maximum catalytic activity was obtained when 3.0% agar-agar was utilized. Both free and entrapped forms the protease was stable at pH-5.0. While, an increase in reaction temperature and time was recorded from 50 to 55°C and 15 to 20 min, respectively. K_M increased from 1.883 mM to 2.399 mM and V_{max} dropped from 1753 U mg^{-1} to 1372 U mg^{-1} upon agar-agar entrapment of enzyme as compared to the soluble form of protease. Moreover, the entrapped enzyme showed a sharp increase in the thermal stability at various temperatures and retained approximately 68.0% of its residual activity at 60°C. However, at this extreme temperature, the soluble protease lost its catalytic performance. Besides, storage sustainability considerably increased as entrapped protease revealed enzymatic activity up to 30 days as compared to the soluble enzyme. Reusability was estimated up to eight cycles (Sattar et al. 2018).

Lactic-co-glycolic acid (PLGA) is a nontoxic, biocompatible synthetic polymer which has been frequently used as a support for enzyme immobilization. It provides better stability, viable uninterrupted operation, and better catalyst reusability; which per se decrease the operational expenses for different industrial processes.

It is shown that biocatalysis properties of the enzyme have improved drastically by immobilization on Au nanoparticles. Also, immobilization of enzymes on chitosan-coated magnetic nanoparticles by adsorption or covalent binding has been demonstrated to increase reusability of them.

Nanofibres, widely used for enzyme immobilization, because of their high specific surface area for high enzyme loading, a porous structure suitable for enzymes active site convenience, low diffusion resistance, easy recoverability and applicable for continuous operations.

Carbon nanotubes (CNTs) have obtained significant research attention in enzyme immobilization. CNTs typically have two types: single-walled carbon nanotubes and multi-walled carbon nanotubes with the diameter around one to tens of nanometers and a length of up to several hundred micrometers (Andrade et al. 2010).

MICROENVIRONMENTS

Over the past 30 years, the portion of immobilized enzymes in industry and market has sharply dropped, because of activity losses and the limitation immobilization procedure themselves. Decreased costs of protein production per se have partially ineffective the advantage of immobilization, that of enzyme recovery and recycling. Instead, microenvironment engineering techniques avoid many of the challenges with immobilization.

One of the more recent trends in enzyme manipulation in order to promote optimal conditions of the protein is controlling the nano- and microscale environment of an enzyme.

Importantly, this new trend of protein engineering for modifying enzymes and control the local environment can be applied simultaneously with traditional enzyme engineering such as rational design and directed evolution of the active site.

A recent example is using a protein microenvironment, or cage-forming protein, to construct an electrostatic barrier between the enzyme and its charged substrates. This nano-cage assembles into a porous icosahedral structure interspersed through the shell. The pores are adequately sized such that small globular proteins and substrates can cross the barrier, which through a series of mutations maintains an overall negative charge (Lancaster et al. 2018).

In another example proteases specificity was altered by peptide fusion. Before being fused to GFP (+36)-TEVp enzyme showed no preference for any of the substrates. Fusion of GFP(+36) to TEVp began to shift substrate selectivity towards the anionic peptide; electrostatic attraction of the peptide resulted in a four times decrease in K_{Mapp}. When assembled in the nanocage, cleavage of the neutral peptide by GFP(+36)–TEVp was reduced by 6-fold, activity toward the zwitterionic substrate was reduced by 7-fold, and the anionic peptide by 100-fold (Lancaster et al. 2018).

In studies directed by Rotello and coworkers, they use gold nanoparticles that functionalized with amino acid to create an environment to Chymotrypsin (ChT) which is charged. By bounding ChT to particles functionalized with glutamic acid residues, they recorded three times increased activity towards cationic substrates. Activity halved with neutral substrates, and the enzyme lost 95% of its activity toward anionic substrate activity. They concluded that the glutamic acid modifications influenced the diffusion of substrate molecules to the complex of nanoparticle-protease. Positively charged substrates could easily access the immobilized enzyme, while diffusion of negatively charged peptides in the local environment was significantly prohibited (You et al. 2005).

CONCLUSION

As proteases are the most utilized enzymes at industrial levels, the emerging technologies for the development of novel proteases which their properties are near the industrial requirements have been reviewed. Moreover, promising molecular biology and protein engineering techniques and progress in material science, nanotechnology and protein chemistries, improved properties such as specificity, catalytic activity, and stability of proteases for broad applications including analytical, biotechnological and therapeutics.

Therefore, proteases may surprise us with the future full of new applications in food and cleaning technologies, pharmaceuticals, fine chemical industries and so on.

REFERENCES

Ahmed, R., Getachew, A. T., Cho, Y.-J., C., Chu, B.-S. (2018). Application of Bacterial Collagenolytic Proteases for the Extraction of Type I Collagen from the Skin of Bigeye Tuna (*Thunnus Obesus*). *LWT-Food Science and Technology*, 89, 44-51.

Andrade, Â. L., Souza D. M., Pereira, M. C., Fabris, J. D., Domingues, R. Z. (2010). PH Effect on the Synthesis of Magnetite Nanoparticles by the Chemical Reduction-Precipitation Method. *Química Nova*, 33 (3), 524-27.

Antonyan, A., Schlenzig, D., Schilling, S., Naumann, M., Sharoyan, S., Mardanyan, S., Demuth, H.-U. (2018). Concerted Action of Dipeptidyl Peptidase IV and Glutaminyl Cyclase Results in Formation of Pyroglutamate-Modified Amyloid Peptides *In vitro*. *Neurochemistry International*, 113, 112-19.

Asgher, M., Bashir, F., Iqbal, H. M. N. (2018). Protease-Based Cross-Linked Enzyme Aggregates with Improved Catalytic Stability, Silver Removal, and Dehairing Potentials. *International Journal of Biological Macromolecules*, 118, 1247-56.

Contesini, F. J., Melo, R. R., Sato, H. H. (2018). An Overview of *Bacillus* Proteases: From Production to Application. *Critical Reviews in Biotechnology*, 38 (3), 321-34.

Diyanat, S., Homaei, A., Mosaddegh, E. (2018). Immobilization of *Penaeus Vannamei* Protease on Zno Nanoparticles for Long-Term Use. *International Journal of Biological Macromolecules*, 118, 92-98.

Duraikannu, D., Chandrasekaran, S. D. (2018). Optimization and Modeling Studies on the Production of a New Fibrinolytic Protease Using Streptomyces Radiopugnans_Vitsd8. *Frontiers in Biology*, 13 (1), 70-77.

Gale, A. J., Miampamba, M., Liu, J., Gonzalez, J. E. (2018). Protease-Activatable Cell-Penetrating Peptides as Prodrugs for Delivery of Cytotoxins into Tumors. *Journal of Clinical Oncology*, 36 (15), e14538-e14538.

Hammami, A., Bayoudh, A., Abdelhedi, O., Nasri, M. (2018). Low-Cost Culture Medium for the Production of Proteases by *Bacillus Mojavensis* Sa and Their Potential Use for the Preparation of Antioxidant Protein Hydrolysate from Meat Sausage by-Products. *Annals of Microbiology*, 68 (8), 473-84.

Hammami, A., Fakhfakh, N., Abdelhedi, O., Nasri, M., Bayoudh, A. (2018). Proteolytic and Amylolytic Enzymes from a Newly Isolated *Bacillus Mojavensis* Sa: Characterization and Applications as Laundry Detergent Additive and in Leather Processing. *International Journal of Biological Macromolecules*, 108, 56-68.

Ibrahim, A. S., Al-Salamah, A. A., El-Toni, A. M., Almaary, K. S., El-Tayeb, M. A., Elbadawi, Y. B., Antranikian, G. (2016). Enhancement of Alkaline Protease Activity and Stability Via Covalent Immobilization onto Hollow Core-Mesoporous Shell Silica Nanospheres. *International Journal of Molecular Sciences*, 17 (2), 184.

Lancaster, L., Abdallah, W., Banta, S., Wheeldon, I. (2018). Engineering Enzyme Microenvironments for Enhanced Biocatalysis. *Chemical Society Reviews*, 47 (14), 5177-5186.

Lau, Y. K., Baytshtok, V., Howard, T. A., Fiala, B. M., Morgan, J. M., Carter, L. P., Baker, D., Lima, C. D., Bahl, C. D. (2018). Discovery and Engineering of Enhanced Sumo Protease Enzymes. *Journal of Biological Chemistry*, 293 (34), 13224-13233.

Marathe, S. K., Vashistht, M. A, Prashanth, A., Parveen, N., Chakraborty, S., Nair, S. S. (2018). Isolation, Partial Purification, Biochemical Characterization and Detergent Compatibility of Alkaline Protease Produced by *Bacillus Subtilis*, *Alcaligenes Faecalis* and *Pseudomonas Aeruginosa* Obtained from Sea Water Samples. *Journal of Genetic Engineering and Biotechnology*, 16 (1), 39-46.

More, S. V., Chavan, S., Prabhune, A. A. (2018). Silk Degumming and Utilization of Silk Sericin by Hydrolysis Using Alkaline Protease from *Beauveria* Sp. (Mtcc 5184): A Green Approach. *Journal of Natural Fibers*, 15 (3), 373-83.

Mótyán, János András, Ferenc, Tóth, József, Tőzsér. (2013). Research Applications of Proteolytic Enzymes in Molecular Biology. *Biomolecules*, 3 (4), 923-942.

Nguyen, M. T. N., Shema, G., Zahedi, R. P., Verhelst S. H. L. (2018). Protease Specificity Profiling in a Pipet Tip Using "Charge-Synchronized" Proteome-Derived Peptide Libraries. *Journal of Proteome Research*, 17 (5), 1923-1933.

Packer, M. S., Rees, H. A., Liu, D. R. (2017). Phage-Assisted Continuous Evolution of Proteases with Altered Substrate Specificity. *Nature Communications*, 8 (1), 956.

Pethe, M. A., Rubenstein, A. B., Khare, S. D. (2017). Large-Scale Structure-Based Prediction and Identification of Novel Protease Substrates Using Computational Protein Design. *Journal of Molecular Biology*, 429 (2), 220-236.

Poeta, P., Dias, A. A., Igrejas, G., Silva, V., Bezerra, R., Nunes, C. S. (2018). Selection, Engineering, and Expression of Microbial Enzymes. In *Enzymes in Human and Animal Nutrition: Principles and Perspectives*, edited by C. S. Nunes and V. Kumar, 1-29, Academic Press.

Ramkumar, A., Sivakumar, N., Gujarathi, A. M., Victor, R. (2018). Production of Thermotolerant, Detergent Stable Alkaline Protease Using the Gut Waste of Sardinella Longiceps as a Substrate: Optimization and Characterization. *Scientific Report*, 8 (1), 12442.

Rigoldi, F., Donini, S., Redaelli, A., Parisini, E., Gautieri, A. (2018). Engineering of Thermostable Enzymes for Industrial Applications. *APL Bioengineering*, 2 (1), 011501.

Sattar, H., Aman, A., Qader, S. A. U. (2018). Agar-Agar Immobilization: An Alternative Approach for the Entrapment of Protease to Improve the Catalytic Efficiency, Thermal Stability and Recycling Efficiency. *International Journal of Biological Macromolecules*, 111, 917-22.

Tavano, O. L., Berenguer-Murcia, A., Secundo, F., Fernandez-Lafuente, R. (2018). Biotechnological Applications of Proteases in Food Technology. *Comprehensive Reviews in Food Science and Food Safety*, 17 (2), 412-436.

Varadarajan, N., Rodriguez, S., Hwang, B.-Y., Georgiou, G., Iverson, B. L. (2008). Highly Active and Selective Endopeptidases with Programmed Substrate Specificities. *Nature Chemical Biology*, 4 (5), 290-294.

Waldner, B. J., Kraml, J., Kahler, U., Spinn, A., Schauperl, M., Podewitz, M., Fuchs, J. E., Cruciani, G., Liedl, K. R. (2018). Electrostatic Recognition in Substrate Binding to Serine Proteases. *Journal of Molecular Recognition*, 31 (10), e2727.

Wijma, H. J., Fürst, M. J. L. J., Janssen, D. B. (2018). A Computational Library Design Protocol for Rapid Improvement of Protein Stability: FRESCO. *Methods in Molecular Biology*, 1685, 69-85.

Xiao, H., Bao, Z., Zhao, H. (2015). High Throughput Screening and Selection Methods for Directed Enzyme Evolution." *Industrial & Engineering Chemistry Research*, 54 (16), 4011-4020.

Xu, B.-L., Dai, M., Chen, Y., Meng, D., Wang, Y., Fang, N., Tang, X.-F., Tang, B. (2015). Improving the Thermostability and Activity of a Thermophilic Subtilase by Incorporating Structural Elements of Its Psychrophilic Counterpart. *Applied and Environmental Microbiology*, 81 (18), 6302-6313.

You, C.-C., De, M., Han, G., Rotello, V. M. (2005). Tunable Inhibition and Denaturation of A-Chymotrypsin with Amino Acid-Functionalized Gold Nanoparticles. *Journal of the American Chemical Society*, 127 (37), 12873-12881.

Zeymer, C., Hilvert, D. (2018). Directed Evolution of Protein Catalysts. *Annual Review of Biochemistry*, 87, 131-157.

Zhang J., Yang, S. J., Gonzalez, F., Yang, J., Zhang, Y., He, M., Shastri, N., Murthy, N. (2018). A Peptide-Based Fluorescent Probe Images Eraap Activity in Cells and in High Throughput Assays. *Chemical Communications*, 54 (52), 7215-7218.

In: Proteases: Functions, Mechanisms and Uses ISBN: 978-1-53615-854-0
Editor: Nelson Pérez Guerra

Chapter 6

Sweet Potato Protein Hydrolysates: Preparation by Enzymatic Hydrolysis, Functional Properties, Antioxidant and Antihypertensive Activities

Tai-Hua Mu*[*] and Miao Zhang
Institute of Food Science and Technology,
Chinese Academy of Agricultural Sciences;
Key Laboratory of Agro-Products Processing,
Ministry of Agriculture and Rural Affairs,
Beijing, China

Abstract

This chapter introduces the overview, preparation and functional properties of hydrolysates from sweet potato protein (SPPH), as well as its antioxidant and antihypertensive activities. It starts by presenting the overview of SPPH, followed by the preparation of SPPH as affected by hydrolysis condition, single and combined enzymes used. It then explains

---

[*] Corresponding Author's E-mail: mutaihuacaas@126.com.

functional properties of SPPH as affected by pH, salts and high hydrostatic pressure (HHP). Then, antioxidant activity of SPPH is introduced, as well as the identification and characterization of its peptides with antioxidant potential. By the end of the chapter, antihypertensive activity of SPPH is presented *in vitro* and *in vivo* level.

Keywords: sweet potato protein hydrolysates, preparation, functional properties, antioxidant activity, antihypertensive activity

1. OVERVIEW

Sweet potato (*Ipomoea batatas* [L.] Lam) is one of the main food crops in China, which accounts for 90.49 and 63.84% of sweet potato production in Asia and the world respectively (FAOSTAT, 2017). Sweet potato contains about 1.7 – 9.1% crude protein on a dry weight basis (Zhang et al., 2014). Sweet potato protein (SPP) possesses a high content of essential amino acids, thus presents higher nutritive value than most other plant proteins (Mu et al., 2009). As a main source of starch in China, sweet potatoes are mainly used to produce starch and starchy foods. During the starch manufacturing, a large amount of industrial waste would be produced, which showed high chemical oxygen demand (COD) and biochemical oxygen demand (BOD), and contained some useful components, such as proteins, sugars and minerals, but were normally discarded directly. Thus, the recovery of SPP is not only helpful for the full utilization of sweet potato resources, but also can reduce the environmental pollution caused by industrial waste during starch manufacturing. At the same time, sweet potato protein hydrolysates (SPPH) have been reported to present remarkable bioactivities, such as antioxidant and antihypertensive, etc. Thus, the production of SPPH would be meaningful to further increase the added value of by-products from starch manufacturing. In this chapter, the recovery of SPP, preparation, functional properties, antioxidant activities and antihypertensive activities were introduced.

2. Preparation of Sweet Potato Protein Hydrolysates (SPPH)

The recovery of SPP from industrial waste during starch manufacturing is one of the main problems faced by the sweet potato starch processing industries in China. SPP has been reported to be recovered by several methods that are introduced below. Protein hydrolysates usually present certain bioactivities, which commonly generated from food proteins by enzymatic hydrolysis, gastrointestinal digestion and fermentation. SPPH has been reported to be prepared by enzymatic hydrolysis and *in vitro* gastrointestinal digestion, which are also summarized below.

2.1. SPP Recovery

There are some different technologies for SPP recovery, such as ammonium sulphate precipitation method, isoelectric precipitation method, membrane filtration, foam fractionation technology and thermal steam technology. All the SPP recovery technologies mentioned above are used the wastewater from sweet potato starch processing as the raw materials, and the process are as follow:

1. Ammonium sulphate precipitation method: fresh sweet potato tuberous roots, washed, color protected by $NaHSO_3$ solution, ground, centrifuged to get the supernatant (wastewater), subjected to 60% saturation of ammonium sulphate saturated buffer, set aside, centrifuged again to get the precipitate (the protein), dissolved the precipitate, and lyophilized to get the SPP powder, then purified by Sephadex G-75 column (Xiong et al., 2009).
2. Isoelectric precipitation method: fresh sweet potato tuberous roots, washed, color protected by $NaHSO_3$ solution at solid to solvent ratio of 1:2, ground, sieved with cheese cloth and centrifuged to get the supernatant (wastewater), adjust the pH of the supernatant to about

4.0 using 2.0 M HCl, stirred and allowed to stand for 85 min at room temperature, centrifuged again to get the precipitate, reconstituted the precipitate in deionized water twice at a protein to solvent ratio of 1:3, readjusted to pH 4.0 with 2.0 M HCl, centrifuged, reconstituted in deionized water at a solid to solvent ratio of 1:3, adjusted the pH to 7.0 with 2 M NaOH, lyophilized to get the SPP powder, of which the protein purity was 62.9% on a dry weight basis determined by Kjeldahl method (Arogundade et al., 2012).

3. Ultrafiltration/Diafiltration: fresh sweet potato tuberous roots, washed, color protected by $NaHSO_3$ solution at solid to solvent ratio 1:2, ground, sieved with cheese cloth and centrifuged to get the supernatant (wastewater), adjusted the pH to 7 using 2.0 M NaOH, ultrafiltered by using 10 kDa molecular weight (MW) cut-off membrane to get the retentate, diluted the retentate 5-fold twice, resuccessively concentrated at a volume concentration factor of 2.5 and 10 respectively, then lyophilized to get the SPP powder, of which the protein purity was 82.1% on a dry weight basis determined by Kjeldahl method (Arogundade et al., 2012).
4. Foam fractionation technology: fresh sweet potato tuberous roots, washed, color protected by $NaHSO_3$ solution (10 mg/mL), ground, centrifuged to get the supernatant (wastewater), subjected to foam separation by an inclined foam separation column, then collect the foamate, and then lyophilized to get the SPP powder. The foam separation conditions were optimized as influent protein concentration 4.51 mg/mL, pH 4.0, air flow rate 0.15 mL/min, influent volume 500 mL, foaming time 100 min, and inclined column angle 30°, by which the protein recovery percentage and enrichment ratios were 84.1% and 1.3, respectively. Compared to the original influent, the biochemical oxygen demand (BOD_5) and chemical oxygen demand (COD_{Cr}) of the residual solution after SPP recovery decreased significantly from 4720 and 8500 mg/mL to 4100 and 7650 mg/mL respectively (Mu et al., 2014).
5. Thermal steam technology: fresh sweet potato tuberous roots, washed, color protected by $NaHSO_3$ solution, ground, subjected to

horizontal screw decanter centrifuge to remove starch and get the supernatant (wastewater), subjected to preheating unit at 45 - 55°C, placed into steam heating flocculation unit (> 80°C) to get protein sedimentation, and then subjected to dewatering and drying unit to get the SPP powder, of which the protein purity was more than 60% on a dry weight basis determined by Kjeldahl method (Mu et al., 2011).

SPP recovered by different methods may have significantly differences on the structural, physicochemical and functional properties, thus affect the production of SPPH by enzymatic hydrolysis.

2.2 Preparation of SPPH by Enzymatic Hydrolysis

Protein hydrolysates are commonly generated from food proteins by enzymatic hydrolysis, which present certain bioactivities that can be used in health care (Valdez-Flores et al., 2016).

The reports on preparation of SPPH by enzymatic hydrolysis were summarized and shown in Table 1. SPPH with acceptable nutritive value were prepared by three single enzymes [Alcalase (ALC), Neutrase (NEU), Protamex (PRO)], and the combinations of enzyme systems [ALC+Flavourzyme (FLA), NEU+FLA, PRO+FLA]. All SPPH presented different degree of hydrolysis (DH), of which the DH prepared by ALC, NEU and PRO were 26.52, 24.97 and 23.84% respectively, while the DH increased to 69.25, 61.32 and 58.67% for ALC+FLA, NEU+FLA and PRO+FLA respectively (Zhang et al., 2012b). Among the three single enzymes, SPPH by ALC presented the highest DH (Zhang et al., 2012b), which might be due to that ALC had a broad specificity on basic (Lys), acidic (Glu), aromatic (Phe, Trp, and Tyr), aliphatic (Leu and Ala), sulphur-containing (Met), and hydroxyl (Ser) amino acid residues (Doucet et al., 2003). FLA contained both endo- and exo-protease activities, which presented higher DHs in sequential hydrolysis by combined enzymes than those single enzymes (Pedroche et al., 2002). SPPH were also prepared by

ALC, proleather FG-F, AS1.398, NEU, Papain and Pepsin, and that by ALC showed the highest DH, followed by Proleather FG-F, AS1.398 and NEU, while papain and pepsin showed relatively poor hydrolysis capacities on SPP with much lower DH (Zhang et al., 2012a). The possible reason was that ALC had a broad specificity, while papain and pepsin presented relatively limited preferential cleavage sites (Doucet et al., 2003; Penas et al., 2006; Ruan et al., 2010).

Response surface methodology (RSM) has been widely applied for optimizing enzymatic hydrolysis conditions among the various statistical experimental design techniques, which presents significant impact on the hydrolysis process of food proteins and their bioactive activities of the resulted hydrolysates. A Box–Behnken design (BBD) in RSM with three independent factors of enzyme-to-substrate ratio (E/S), pH value and reaction temperature at three variation levels was applied to optimize the production of antioxidant hydrolysates from SPP by ALC. Total 17 experiments including 12 factorial points and 5 replicates at the central point were designed, and ·OH radical scavenging activity and Fe^{2+}-chelating activity were selected as the responses (Zhang and Mu, 2016). The predicted quadratic regression models were obtained after application of the response surface regression procedure, of which the determination coefficient (R^2) was 0.9967 and 0.9915 for ·OH radical scavenging activity and Fe^{2+}-chelating activity respectively, indicating that 99.67 and 99.15% of the total variations for ·OH radical scavenging activity and Fe^{2+}-chelating activity were attributed to the independent variables respectively. The predicted R^2 values were of 0.9924 and 0.9769, indicating that the agreements were satisfactory between ·OH radical scavenging activity, Fe^{2+}-chelating activity predicted by the model and the experimental data respectively. At the same time, the Fisher's F-test showed a high F-value of 71.00 and 173.13, and very low *P*-value of < 0.0001 and < 0.0001 for ·OH radical scavenging activity and Fe^{2+}-chelating activity respectively, and the lack of fit was not significant for both ($P = 0.45 > 0.05$; $P = 0.63 > 0.05$), which would give good fits to the mathematic models.

Table 1. Preparation of SPPH by enzymatic hydrolysis

No.	Enzyme used	Bioactivity	References
1	Single enzymes (ALC, NEU and PRO), and combinations of enzyme systems [ALC + FLA, NEU + FLA, PRO + FLA]	superoxide, hydroxyl and DPPH radical scavenging activities	Zhang et al., 2012b
2	ALC, proleather FG-F, AS1.398, NEU, Papain and Pepsin	·OH scavenging activity; Fe^{2+}-chelating activity; protective effect against oxidative DNA damage	Zhang et al., 2012a
3	ALC[a]	·OH scavenging effect; Fe^{2+}-chelating activity;	Zhang and Mu, 2016
4	ALC[b]	·OH scavenging activity; Fe^{2+}-chelating ability; Oxygen radical absorbance capacity (ORAC)	Zhang and Mu, 2017
5	Pepsin	Angiotensin I-converting enzyme (ACE) inhibitory activity (*in vitro* and *in vivo*)	Wang et al., 2011
6	Orientase 22BF, Orientase ONS, Nucleisin, Orientase 20A, Protin FN, Protin AC10F, Glutaminase Daiwa C100S, Protin PC 10F, Thermoase PC 10F, Protease A, Protease M, Protease N, Protease P, Protease S, Protease F, Proleather FG-F	Angiotensin I-converting enzyme (ACE) inhibitory activity (*in vitro* and *in vivo*)	Ishiguro et al., 2012
7	Thermoase PC10F, Protease S and Proleather FG-F	lipid metabolism in mice administered a high-fat diet	Ishiguro et al., 2016

[a] Optimization by response surface methodology;

[b] Effect of high hydrostatic pressure (HHP).

Three-dimensional response surface plots, the graphical representations of regression equations, could reflect the interaction of two variables when keeping other independent variables at the 0 level. The ·OH scavenging activity of SPPH firstly increased rapidly with increasing E/S, pH value and reaction temperature, then slowed down steadily with the further increasing of reaction temperature, and the similar trends were found in the response surface plots of Fe^{2+}-chelating activity. The optimum parameters for SPPH preparation was obtained by employing the Design-Expert software, which were E/S ratio 4%, reaction temperature 57°C and pH value 7.8 (Zhang and Mu, 2016).

High hydrostatic pressure (HHP) can offer novel tastes and textures of foods, and has been utilized realistically in food industry after the significant development over the last 20 years (Norton and Sun, 2008). HHP technology could help improve digestibility and enzymatic hydrolysis process of food protein, and improved the bioactivities of the resulted hydrolysates. HHP could also change the conformations and extend the molecular chains of proteins, which would expose new restriction sites and thus were beneficial for the enzymatic hydrolysis process (Bonomi et al., 2003). It was reported that DH of whey proteins hydrolysates increased significantly with HHP at 100 – 400 MPa for 5 – 30 min assisted enzymatic hydrolysis (Ambrosi et al., 2016). SPPH were produced by ALC under HHP at 100, 200 and 300 MPa for 30 and 60 min respectively, of which the DH was significantly different ($P < 0.05$) (Zhang and Mu, 2017).

Under atmospheric pressure (0.1 MPa), the DH of SPPH by ALC was 20.83 and 23.17% with hydrolysis time for 30 and 60 min respectively. DH of SPPH by ALC were significantly improved by HHP at applied pressure from 100 to 300 MPa and hydrolysis time from 30 to 60 min, which were 22.65, 23.34 and 25.78% under 100, 200 and 300 MPa for 30 min, and reached up to 24.56, 25.26 and 31.68% under 100, 200 and 300 MPa for 60 min respectively (Zhang and Mu, 2017).

In addition, SPPH with angiotensin I-converting enzyme (ACE) inhibitory activity were prepared with pepsin. The effect of substrate concentration, E/S, pH value and temperature on the ACE inhibitory activity was studied by one- factor experiment and optimum design of quaternary quadratic rotation orthogonality experiment to get the best enzymatic hydrolysis condition.

The optimal processing conditions were as follows: substrate concentration 2.3%, E/S 3.7%, pH 2.3, temperature 37°C and time 8 h

(Wang et al., 2011). It was also reported that 16 commercial proteases (Orientase 22BF, Orientase ONS, Nucleisin, Orientase 20A, Protin FN, Protin AC10F, Glutaminase Daiwa C100S, Protin PC 10F, Thermoase PC 10F, Protease A, Protease M, Protease N, Protease P, Protease S, Protease F, Proleather FG-F) were used for preparation and selection of SPPH with high ACE inhibitory activity, as well as the combination of two or three proteases selected above.

The enzymatic reaction was performed with substrate concentration 1%, hydrolysis time 16 h, and optimum pH, temperature and E/S (Ishiguro et al., 2012). SPPH prepared by Orientase 22BF, Thermoase PC 10F, Protease P and Protease F showed high protein digestibilities, and the combination of the four protease showed higher digestibilities, which ranged from 76.0% (Orientase 22BF + Protease F) to 95.0% (Thermoase PC 10F + Protease P + Protease F).

Furthermore, three commercial proteases (Thermoase PC10F, Protease S and Proleather FG-F) were used to prepare SPPH, of which the effect on lipid metabolism in mice administered a high-fat diet was determined (Ishiguro et al., 2016).

The enzymatic hydrolysis condition was as follows: precipitate 6.0 kg, Thermoase PC10F 1.5 g, Protease S 15 g, Proleather FG-F 15 g, pH 8.5, temperature 65°C and hydrolysis time 16 h.

2.3. Preparation of SPPH by Gastrointestinal Digestion

SPPH were prepared from the major root storage proteins Trypsin inhibitors (TIs) of sweet potato by *in vitro* digestion using pepsin treatment or sequential pepsin and chymotrypsin treatment. The condition of digestion was as follows: for pepsin treatment, TIs 4 mg, pepsin 0.8 mg, pH 2.0, reaction time 8 h, temperature 37°C; then for continuous chymotrypsin treatment, chymotrypsin 0.8 g, pH 8.3, reaction time 4, 8 and 12 h. It was showed that the hydrolysis by pepsin treatment or sequential treatment of pepsin and chymotrypsin resulted significantly increase in free amino groups of TIs (Hou et al., 2005). Zhang and Mu (2016) prepared SPPH by ALC, then the obtained SPPH were further subjected to simulated *in vitro* gastrointestinal digestion (GID) by sequential hydrolysis using pepsin and pancreatin. The condition for GID was as follows: for simulated gastric digestion (GD) by pepsin, E/S 4:100 (w/w), pH 1.5, temperature 37°C, reaction time 60 and 120 min; for simulated GID, E/S 4:100 (w/w), pH 7.8, temperature 40°C, reaction time 120 min. Compared with SPPH before GID, the MW > 3 kDa peptide fractions in SPPH were decreased from 30.56 to 20.05%, whereas the MW < 3 kDa peptides were significantly increased from 69.44 to 79.95% (Zhang and Mu, 2016). In addition, Zhang and Mu (2019) investigated the effects of thermal (boiling, steaming and autoclaving), microwave and ultrasonication pretreatments on the production of SPPH through GID, and the results showed that all pretreatments significantly increased the DH, antioxidant activities and MW < 3 kDa peptide fractions contents of SPPH by GID, and diverse peptides ranged from 487.24 to 1477.74 Da with 7 – 13 amino acids were identified.

3. Functional Properties of SPPH

SPPH could be prepared by enzymatic hydrolysis and *in vitro* GID, which showed noteworthy antioxidant activity and great protective effect on oxidative DNA damage (Zhang and Mu, 2014; Zhang and Mu, 2016; Zhang and Mu, 2019). However, few studies on the functional properties of SPPH are presently available. As the emulsifying property is one of the most important functional properties of food proteins and its hydrolysates, it would be necessary to evaluate the emulsifying properties of SPPH as affected by different environmental factors (HHP treatment, peptide concentration and oil volume fraction, pH and salts) that are summarized below.

3.1. Emulsifying Properties of SPPH Affected by HHP

SPPH was prepared with five different enzymes (Papain, AS1.398, NEU, ALC and Proteinase K) under 0.1, 100, 200, 300 and 400 MPa for 2, 4, 6, 8 and 10 min as reported by Cui (2014).

Table 2. Preparation of SPPH by gastrointestinal digestion

No.	Enzyme used	Bioactivity	References
1	Pepsin; sequential treatment of pepsin and chymotrypsin	DPPH radical scavenging activity	Hou et al., 2005
2	Pepsin, Pancreatin	·OH scavenging activity; Fe^{2+}-chelating activity	Zhang and Mu, 2016
3	Pepsin, Pancreatin	·OH scavenging activity; Fe^{2+}-chelating activity; ORAC	Zhang and Mu, 2019

At atmospheric pressure (0.1 MPa), the emulsion of SPPH prepared by all five enzymes showed large and sparse droplets, and the emulsion droplets aggregated rapidly after homogenization. Compared to the emulsion at 0.1 MPa, the emulsion droplets of SPPH by ALC under HHP were much sparser; those by AS1.398 and NEU were denser and smaller; those by proteinase K were more uniform at 100 and 300 MPa; while those by Papain were more dense and homogeneous.

At 0.1 MPa, emulsifying activity index (EAI) of SPPH by five enzymes was ranging from 16.43 to 40.69 m^2/g. Under HHP, EAI of SPPH by ALC increased first and then decreased with pressure increasing, but was significantly lower than that under 0.1 MPa; those by AS1.398 and NEU increased significantly and reached high values under 300 (67.98 m^2/g) and 400 (44.42 m^2/g) MPa respectively; those by Papain increased significantly being was 101.59 m^2/g at 300 MPa; while those by proteinase K was also significantly higher than that at 0.1 MPa.

3.2. Emulsifying Properties of SPPH Affected by Peptide Concentration and Oil Volume Fraction

SPPH by Papain under 300 MPa for 6 min was prepared, and peptide concentrations of 0.10, 0.25, 0.50, 0.75 and 1.00% (w/v) and oil volume fractions of 5, 15, 25, 35 and 45% were used to determine their effects on the emulsifying properties of SPPH (Cui, 2014).

For the microstructure of SPPH emulsion, the emulsion droplets were fine and uniformly distributed without aggregation at a peptide concentration of 0.10%, of which the size increased significantly at peptide concentrations of 0.25 - 1.00% showing a non-uniform state. With the increase of peptide concentration, the emulsion droplets began to aggregate, especially when the concentration reached up to 0.75%. With the increase of the oil volume fraction, the density of emulsion droplets increased obviously, and the size gradually became uniform and small.

For emulsion particle size ($d_{4,3}$), it was 50.06 μm at peptide concentration of 0.10%, which decreased gradually with the increase of peptide concentrations and was 41.89 μm at peptide concentration of 1.00%. With the increase of the oil volume fraction, the $d_{4,3}$ of emulsion droplets decreased from 62.73 μm at 5% to 49.81 μm at 45%, indicating that the larger the oil volume fraction, the smaller the $d_{4,3}$ of emulsion droplets.

For the EAI, it was decreased significantly with the increase of peptide concentrations. At peptide concentration of 0.10%, the EAI was the highest, which was 99.81 m^2/g. When peptide concentration reached to 1.00%, the EAI of SPPH emulsion decreased to 88.79 m^2/g. The EAI of SPPH emulsion increased significantly with the increase of oil volume fraction. When oil volume fraction was just 5%, the EAI was 25.33 m^2/g; when oil volume fraction was 45%, the EAI was the highest, being 102.87 m^2/g.

For the emulsifying stability index (ESI), it was also decreased gradually with the increase of peptide concentrations. When peptide concentrations increased from 0.10 to 1.00%, ESI decreased from 21.15 to 13.49 min. At 5% oil volume fraction, ESI of SPPH emulsion was 30.41 min, which decreased to 17.83 to 22.26 min at 15 - 45% oil volume fraction.

For the apparent viscosity of SPPH emulsion, it was decreased gradually with the increase of shear rate within the set shear rate range (1-100 s^{-1}), showing different degrees of shear thinning. At 0.10% peptide concentration, the initial apparent viscosity of SPPH emulsion was the highest, which was 77.10 mPa·s. Compared to those at 5 - 35% oil volume fraction, the initial apparent viscosity of SPPH emulsion at 45% was the highest, which was 1165 mPa·s.

The above results showed that peptide concentrations and oil volume fraction significantly affected the emulsification characteristics of SPPH. The $d_{4,3}$ and apparent viscosity of SPPH emulsion were closely related to the EAI and ESI. The increase in peptide concentrations led to the decrease of $d_{4,3}$, the initial apparent viscosity, EAI and ESI. The increase in oil volume fraction would lead to the decrease of $d_{4,3}$, the increase of the initial apparent viscosity, EAI and ESI.

3.3. Emulsifying Properties of SPPH Affected by pH and Salts

The stability of an emulsion is affected by pH value and ion concentration (such as sodium chloride, calcium chloride, etc.). The change of pH value can change the charge of emulsion droplets, which lead to different repulsion and attraction between the droplets, and then affects the flocculation degree and the droplets particle size. Ions may cross-link and denaturant with emulsifiers, change their molecular structure, resulting in the decrease of emulsifying ability of emulsifiers, thus making the emulsion unstable, and may also change the structure of the emulsified object and affect the stability of the emulsion.

SPPH by Papain under 300 MPa for 6 min was prepared. Effects of different pH values (3, 5, 7 and 9) and NaCl concentrations (0.2, 0.4, 0.6, 0.8 and 1.0 M) on the emulsifying properties of SPPH were determined by Cui (2014).

For the microstructure of SPPH emulsion, with the pH values increasing from 3 to 9, the particle size of emulsion droplets decreased obviously, while the density and homogeneity increased obviously. The emulsion droplets of

SPPH at low NaCl concentrations (0.2 - 0.4 M) were uniform in size, while those at high NaCl concentrations (0.6-1.0 M) showed non-uniform.

For emulsion particle size ($d_{4,3}$), it was 54.19 μm at pH 7 without NaCl addition, which was decreased from 64.45 to 51.21 μm with the pH value from 3 to 9. The $d_{4,3}$ was increased to 59.70 μm after adding 0.2 M NaCl. With the increase of NaCl concentrations, $d_{4,3}$ showed a significant increasing trend. When NaCl concentration up to 1.0 M, $d_{4,3}$ reached the highest value, which was 69.72 μm.

For the EAI, it was increased significantly with the pH value increased from 3 to 9. EAI of SPPH was 56.35 m^2/g at 0.2 M NaCl, which was decreased significantly with the increase of NaCl concentrations. For the emulsifying stability index (ESI), it was 14.84 min at pH 7 without NaCl addition. The ESI of SPPH was 29.13 min at pH 3, which decreased significantly with the increase of pH values (5-9). The ESI of SPPH was 13.19 min at 0.2 M NaCl, which was increased first and then decreased with the increase of NaCl concentrations.

The apparent viscosity of SPPH emulsion with different pH values and NaCl concentrations were decreased gradually with the increase of shear rate within the set shear rate range (1 - 100 s^{-1}), showing different degrees of shear thinning. With the increase of shear rate, the apparent viscosity of SPPH emulsion decreased gradually, indicating that the flocculation occurred. The apparent viscosity of SPPH emulsion at pH 7 without NaCl addition was 82.85 mPa·s, which increased to 112.00 mPa·s at pH 3, but decreased to 54.90 mPa·s at pH 9.0.

The above results showed that pH values and NaCl concentrations significantly affected the emulsification characteristics of SPPH. The pH values could affect the emulsifying properties of SPPH by changing the surface charge distribution. Under acidic conditions (pH 3), the EAI of SPPH was the lowest, but the ESI was the highest. The increasing in the pH values effectively improved the emulsifying activity of SPPH. Emulsifying properties of SPPH could be improved at low NaCl concentration (0.2 M), which would be decreased at high NaCl concentration (0.4 - 1.0 M).

4. Antioxidant Activity of SPPH

Reactive oxygen species (ROS) are produced by all aerobic cells that have two subgroups: free radicals (superoxide radical, $O^{2.-}$; hydroxyl radical, ·OH) and non-radical ROS (hydrogen peroxide, H_2O_2) (Sarsour et al., 2009; Dayem et al., 2010). In normal cells, through balancing oxidants (ROS) by antioxidants during metabolism and other activities, the redox homeostasis can be maintained (Smith et al., 2011). However, when the level of ROS exceeds the tolerance capacity for antioxidant defense system, cellular macromolecules, such as fatty acids, proteins and DNA, will be damaged by ROS-mediated oxidative stress, which could subsequently lead to gene mutation and further chronic diseases development (Shackelford et al., 2009).

Peptides derived from various food proteins by enzymatic hydrolysis have emerged as a new source of natural antioxidants, which possess acceptable nutritive value without marked adverse effects and help to reduce the risks of ROS-related diseases of the human body. Antioxidant peptides mainly exert their activity by hydrogen transfer and electron donation, which present a great potential to be utilized for food and pharmacological applications (Lorenzo et al., 2018). Thus, it would be important to evaluate the antioxidant activity of SPPH, which are summarized below.

4.1. Antioxidant Activity

SPPH from TIs prepared by pepsin or sequential pepsin and chymotrypsin treatment showed increased DPPH scavenging activity with the increasing treatment time, which might be attributed to the increased free amino groups and inner region of amino acids of TIs after enzymatic hydrolysis (Hou et al., 2005).

SPPH prepared with single enzymes (ALC, NEU and PRO), and combinations of enzyme systems [ALC+FLA, NEU+FLA, PRO+FLA] presented significant increases in free radical scavenging activities compared with SPP. Compared to other hydrolysates, ALC hydrolysates exhibited the highest superoxide, ·OH and DPPH radical scavenging activities, which was 18.71, 27.13 and 90.10% respectively. The possible reason was that the contents of several antioxidant amino acid residues (His, Met, Tyr and Phe) in ALC hydrolysates were much higher than those in SPP and other hydrolysates (Zhang et al., 2012b).

SPPH prepared by six enzymes (ALC, Proleather FG-F, AS1.398, NEU, Papain and Pepsin) showed different antioxidant activities, among which, ALC hydrolysates exhibited the highest ·OH scavenging activity and Fe^{2+}-chelating ability with IC_{50} 1.74 and 1.54 mg/ml respectively. In addition, the MW < 3 kDa fractions of SPPH by ALC showed the highest antioxidant activities, which was probably because of the increase of several antioxidant amino acids, such as His, Met, Cys, Tyr and Phe, as well as the hydrophobic amino acids (Zhang et al., 2012a).

SPPH were also generated by ALC under HHP at 100, 200 and 300 MPa. HHP significantly improved the antioxidant activity and increased the MW < 3 kDa fraction content. SPPH by ALC under 200 and 300 MPa for 60 min showed the highest ·OH scavenging activity, which were 56.21 and 54.93% respectively; those under 100 and 200 MPa for 60 min showed the highest Fe^{2+}-chelating ability, which were 95.91 and 96.16% respectively; whereas those under 0.1 MPa for 30 min showed the highest ORAC value (123.93 μg of TE), which had no significant difference with SPPH under 100 MPa (117.50 μg TE/mL) and 300 MPa (116.67 μg TE/mL) for 60 min (Zhang and Mu, 2017).

In addition, the MW < 3 kDa peptide fractions content in SPPH via GID by Thermal, microwave and ultrasonication pretreatments exhibited positive correlations with its ·OH scavenging activity, Fe^{2+}-chelating capacity and ORAC (Zhang and Mu, 2019).

4.2. Protective Effects on Oxidative DNA Damage

DNA is a very sensitive bio-target of ROS. DNA damage by ROS can serve as an initiation of certain chemical carcinogenesis or feature in the pathogenesis of neurodegenerative disorders (Toyokuni, 1996; Smith et al., 1997).

The protective effect of different MW fractions (> 10, 3 - 10 and < 3 kDa) from SPPH by ALC against DNA damage induced by ·OH and via chelating Fe^{2+} ion was investigated by Zhang et al. (2012a). The protective effects against ·OH-induced DNA damage for MW > 10, 3 - 10 and < 3 kDa fractions were 26.08, 37.38 and 50.87% respectively. When the SPPH fractions prereacted with Fe^{2+} ion, DNA was protected from the cleavage, suggesting that different MW fractions from SPPH by ALC could inhibit the generation of ·OH. The inhibited effect on DNA damage via chelating Fe^{2+} ion for MW > 10, 3 - 10 and < 3 kDa fractions were 47.86, 16.56 and 72.47% respectively. The results above were in accordance with the report by Leonard et al. (2006), who pointed out that DNA damage induced by ROS could be inhibited through two paths: (1) scavenging the radicals produced during oxidation reactions; (2) inhibiting the generation of radicals.

4.3. Purification, Identification and Characterization of Antioxidant Peptides

The purification of antioxidant peptides from SPPH has been reported to carry out by peptide fractionation through ultrafiltration using centrifuge tubes with different MW cut-off membranes, size exclusion chromatography (SEC) followed by semi-preparative reverse-phase high-performance liquid chromatography (RP-HPLC) (Zhang and Mu, 2016).

The identification of antioxidant peptides from SPPH was analyzed by LC-MS/MS and/or LC-QTOF-MS/MS (Zhang and Mu, 2017; Zhang and Mu, 2019). While, peptide conformations were carried out by PEP-FOLD3 (a free online resource) for the characterization antioxidant peptides from SPPH (Zhang and Mu, 2017).

Some amino acids in peptide chains present certain intrinsic properties with potential antioxidant activities, which are also attributed to their positions in the peptide sequences sometimes. It was reported that Trp, Tyr and Met amino acids presented the strongest antioxidant capacities, followed by Phe, His and Cys (Hernández-Ledesma et al., 2005). Strong antioxidant activities of Tyr, Trp and Phe are considered to be due to their hydrogen donation ability; Met is owing to the oxidization to its sulfoxide form; Cys is contributed to the sulfur hydrogen; while His shows a proton donation capability with its imidazole group. Peptide conformations also synergistically or antagonistically affected antioxidant activity of peptides in comparison to amino acids themselves.

Zhang and Mu (2014) indicated that 5 peptides were identified by LC-MS/MS in the antioxidant F-IV fraction by semi-preparative RP-HPLC, of which Tyr-Tyr-Ile-Val-Ser, Thr-Tyr-Gln-Thr-Phe and Ser-Gly-Gln-Tyr-Phe-Leu were from IV-5c, while Tyr-Tyr-Asp-Pro-Leu and Tyr-Met-Val-Ser-Ala-Ile-Trp-Gly were from IV-5i. To confirm their antioxidant activities, the 5 five peptides were synthesized, and their ·OH scavenging activities were 74.52, 70.83, 56.30, 33.28 and 11.61% respectively, suggesting that the peptide Tyr-Tyr-Ile-Val-Ser presented the highest antioxidant activity.

Zhang and Mu (2017) identified 20 peptides with potential antioxidant amino acids from dominant fractions with high ORAC values by semi-preparative RP-HPLC, among which 14 peptides showed certain ORAC ranging from 30.54 to 123.06 μg TE/mL, and peptide HDSASGQY presented the highest ORAC value, followed by peptides YYMVSA, RYYDPL, YYIVS and HDSESGQY. For the 5 peptides with high ORAC values, peptide HDSASGQY showed His (positively charged, antioxidant) at the N1 position, Gly (hydrophobic) in the C3 region, and Tyr (bulky aromatic, antioxidant) at the C1 domain.

Peptide YYMVSA presented two Tyr (bulky aromatic, antioxidant) at the N1, N2 domain, and a Met (bulky hydrophobic, antioxidant) at the C4 region. Peptide RYYDPL presented an Arg (bulky positively charged) at the N1 domain, and two Tyr in the N2, N3 position. Peptide YYIVS also contained two Tyr at the N1, N2 domain, and an Ile (bulky hydrophobic) at the C3 region. In addition, peptide HDSESGQY showed His at the N1 position, Tyr at the C1 domain, and Glu (negatively charged) in the N4 position.

Zhang and Mu (2019) also found 27 diverse peptides from MW < 3 kDa peptide fractions in SPPH by GID with autoclaving pretreatment analyzed by LC-QTOF-MS/MS, of which the MWs were in the range of 487.24 to 1477.74 Da with 7 – 13 amino acids, and their antioxidant activities might be attributed to the presence of well-known antioxidant amino acids of Tyr, Trp, Phe and/or Met both in N and C terminal regions, as well as at the position inside peptide sequences. Within the 27 peptides, peptide YFLKAGEF and YYDPLTRA presented the most folded (4 turns) and compact structures; peptide YFLKAGEF showed 3 aromatic rings (1 from Tyr and 2 from Phe residues) appeared in external positions; whereas peptide YYDPLTRA had 2 aromatic rings of Tyr and 1 hydrophobic Pro presented in external positions.

5. Antihypertensive Activity of SPPH

Hypertension is a health problem worldwide, which is associated with sedentary lifestyle, mainly observed in adults and is treated with drugs that produce side effects. Hypertension refers to persistent raised blood pressure in the arteries and is a long term medical condition initiating major risk factors for Cardio-Vascular Diseases (CVD).

Peptides with antihypertensive acitivity could be an alternative or a supplemental treatment for hypertensive patients. At present, many antihypertensive peptides have been reported, however few studies have been reported to determine their effectiveness *in vivo*. Thus, it would be important to evaluate the antihypertensive activity of SPPH both *in vitro* and *in vivo*, which are summarized below.

5.1. *In vitro* ACE Inhibitory Activity of SPPH

5.1.1. Hydrolysis Time on ACE Inhibitory Activity of SPPH

Wang et al. (2011) reported that ACE inhibitory activity of SPPH significantly increased with the hydrolysis time increasing from 1 to 8 h. At the hydrolysis time 8 h, ACE inhibitory activity of SPPH reached the maximum value of 78.37%. When the hydrolysis time increased up to 12 h, there was no significant change on the ACE inhibitory activity of SPPH. While, ACE inhibitory activity of SPPH significantly decreased when the hydrolysis time continued to increase.

5.1.2. Peptide Concentration and MW on ACE Inhibitory Activity

ACE inhibitory activity of peptides with MW 0 - 3, 3 - 5 and 5 -1 0 kDa from SPPH was studied by Wang (2011). With peptide concentration (0-2.5 mg/ml) increasing, ACE inhibitory activity of peptides with MW 0-3, 3-5 and 5-10 kDa significantly increased, and showed a dose-dependent manner. At the concentration of 2.5 mg/ml, ACE inhibitory activity of peptides with MW 0-3, 3-5 and 5-10 kDa from SPPH reached the maximum values, which were 77.30, 71.94 and 53.62% respectively. Compared with the peptides with MW 3 - 5 and 5 - 10 kDa, ACE inhibitory activity of peptides with MW 0-3 kDa was higher, which showed lower IC_{50} value of 0.67 mg/ml.

5.1.3. MW Distribution and Amino Acid Composition of ACE Inhibitory Peptides from SPPH

MW distribution of ACE inhibitory peptides from SPPH was determined by MALDI-TOF-MS. The MW of 0 - 3 kDa peptides was ranged from 212 to 2551 Da, and the content of peptides with MW of 1384, 1131, 985 and 815 Da is more abundant.

Amino acid composition of ACE inhibitory peptides with MW 0-3 kDa from SPPH was determined by Wang (2011). The content of hydrophobic amino acids (Ala, Ile, Leu, Met, Phe, Pro, Tyr and Val) in MW 0-3 kDa peptides was 29.80 g/100g peptides; the content of polar amino acids with uncharged side chain (Thr, Ser and Cys) was 10.22 g/100 g peptides; that of basic amino acids (Lys, His and Arg) was 5.61 g/100 g peptides; that of acidic amino acids (Glu and Asp) was 14.41 g/100 g peptides; that of aromatic amino acids (Tyr and Phe) was 8.41 g/100 g peptides; while that of branched chain amino acids (Leu, Ile and Val) was 12.94 g/100 g peptides. Ruiz et al. (2004) suggested that the content of amino acids Pro and Leu had a great effect on their ACE inhibitory activity, as well as their position at the

C-terminus of short peptides. Jung et al. (2006) reported the peptides containing a large amount of hydrophobic and aromatic amino acids showed high antihypertensive activity, and ACE inhibitory activity of the peptides would be stronger if these two kinds of amino acids were located at the C-terminus. In addition, the increases of hydrophobic, aromatic and branched chain amino acids contents are beneficial to the increase of ACE inhibitory activity of protein hydrolysates. The contents of hydrophobic aromatic and branched chain amino acids in ACE inhibitory peptides in SPPH were 29.80, 8.41 and 12.94 g/100g peptides respectively, resulting in strong ACE inhibitory activity of SPPH.

5.2. *In Vivo* Antihypertensive Activity of SPPH

It is worth noting that many short peptides have an inhibitory effect on ACE *in vitro*, but no antihypertensive effects are found *in vivo* sometimes. A peptide Tyr-Gly-Gly-Tyr isolated from rice wine had an IC_{50} value of 16.2 μmol/L for ACE inhibitory activity *in vitro*, but no antihypertensive effect presented on an oral dose of 100 mg/kg for spontaneously hypertensive rats (SHRs) (Saito et al., 1994). This might be due to that ACE inhibiting peptides had been degraded into inactive fragments during the gastrointestinal digestion. Another possibility was that these inhibitory peptides were only substrates for ACE, but not true inhibitors, so they had high ACE inhibitory activity by *in vitro* assays but had no effect when determined *in vivo*. In addition, some ACE inhibitory peptides had low activity *in vitro*, but *in vivo* test results showed that they had significant antihypertensive effects. This might be because of that ACE inhibitory peptides were degraded by the digestive enzymes in the gastrointestinal tract after being taken into the human body, and fragments with antihypertensive activity were generated. Therefore, the antihypertensive activity of the polypeptides must be verified by *in vivo* animal testing.

SHRs are internationally recognized as the animal models of hypertension, being one of the best models for studying spontaneous hypertension. Normal rats have a systolic blood pressure (SBP) of 110 - 120 mmHg, whereas the blood pressure of SHR is as high as 200 mmHg. SHRs were orally administered with peptides from peptic digest of wakame for one week, of which the blood pressure decreased 25-35 mmHg and the antihypertensive effect lasted for 3-8 weeks (Suetsuna and Nakano, 2000). Tsai et al. (2008) fermented milk with *Lactobacillus bulgaricus* and *Streptococcus thermophilus* for 5 h at 43°C, and Protease was added at the initial stage of fermentation. After the reaction was terminated, the IC_{50} value of ACE inhibitory activity of the obtained product was 0.27 mg/ml, which was separated into four components by Sephadex G-15, among which the fourth component Tyr-Pro-Tyr-Tyr showed a kinetic mode of mixed inhibition and the IC_{50} value was lowest (90.0 μmol/L). Then, diluted fermented milk whey with peptide concentration 4.9 mg/ml was subjected to the *in vivo* measurement of blood pressure using SHRs. After 8 weeks, the systolic and diastolic blood pressure of the SHR decreased by 15.9 and 15.6 mmHg, respectively. Wang (2011) confirmed the ACE inhibitory activity of SPPH by pepsin, and also verified the antihypertensive effect of it *in vivo* summarized as below.

5.2.1. Group of Test Animals

Ten-week-old female SHRs (120 ± 10 g) were used. All SHRs were kept in single cages and were allowed free access to the diets and water. Animals were kept under a 12 h light/dark cycle, 22 ± 2°C ambient temperature and 55 ± 10% relative humidity for 7 days prior to experimental use. SHRs were randomly divided into 5 groups with 5 rats per group: blank control (Group 1, same volume of distilled water), positive control (Group 2, Captopril 40 mg/kg body weight), high-dose SPPH (Group 3, 2000 mg/kg body weight), middle-dose SPPH (Group 4, 1500 mg/kg body weight) and low-dose SPPH (Group 5, 600 mg/kg body weight).

5.2.2. Effect of Single Administration of SPPH on SBP and Heart Rate of SHR

Compared with the blank control, SPPH with different dose and positive control with captopril pretended antihypertensive activity within 2 to 8 h ($P < 0.05$). The maximum decrease in blood pressure in SHRs with high-, middle- and low-dose SPPH occurred after 4 h administrations, which were 23.59, 16.91 and 16.18 mmHg respectively. Then the blood pressure began to rise, which returned to the original level after 12 h administrations. After oral administration of different doses of SPPH, the heart rate of SHRs in each group was not significantly different from the blank control group within 12 h, thus indicating that SPPH had no adverse effect on the heart rate of SHRs.

5.2.3. Effects of Long-Term Administration of SPPH for 24 Days on SBP and Heart Rate in SHR

SBP of SHRs in positive control group after long-term administration for 24 days decreased significantly with the increasing time. The administration of high-, middle- and low-dose SPPH decreased the SBP of SHRs in different degrees, which decreased 13.01, 10.54 and 8.46 mmHg after 6 days, respectively. Subsequently, the SBP of each group with different dose increased to various degrees with increasing time, which was due to that the SBP increased with the increase of SHR age. Within 24 days, the SBP of each group with different dose was significantly lower than that of the blank group ($P < 0.05$), which indicated that the dynamic change of SBP stably decreased with the long-term administration of SPPH. After oral administration of different doses of SPPH, the heart rate of SHRs in each group was not significantly different from the blank control group within 24 days, thus indicating that SPPH had no adverse effect on the heart rate of SHRs.

5.2.4. Effects of Long-Term Administration of SPPH for 24 Days on Plasma Lipid Level of SHR

The total cholesterol (TC), triacylglycerol (TG), high-density lipoprotein cholesterol (HDL-C) and low-density lipoprotein cholesterol (LDL-C) were measured to determine whether captopril, high-, middle- and low-dose SPPH showed some adverse effects on the plasma lipid level of SHRs. After administration for 24 days, the TC, TG, LDL-C, and HDL-C in the blank control group were 2.37, 1.32, 1.93 and 0.25 mmol/L respectively. For positive control, high-, middle- and low-dose SPPH group, the TC, TG,

HDL-C and LDL-C of SHRs were in the range of 2.10 - 2.2, 1.08 - 1.39, 1.72 - 1.78 and 0.22 - 0.23 mmol/l respectively, showing no significant difference with those of blank control.

5.2.5 Effects of Long-Term Administration of SPPH for 24 Days on Endothelin-1 (ET-1) Content of SHR

ET is one kind of peptides containing 21 amino acid residues, including ET-1, ET-2 and ET-3, in which ET-1 is produced by vascular endothelial cells and epithelial cells, being the strongest vasoconstrictor substance and can cause almost all arteriovenous contractions. Compared with the blank control, the positive control (captopril), high-, middle- and low-dose SPPH groups showed lower ET-1 content. The content of ET-1 for low-dose SPPH group was 15.45 pg/ml, which significantly decreased to 12.79 and 12.6 pg/ml respectively. The results suggested that as the concentration of SPPH increased from 600 to 2000 mg/kg body weight, the serum ET-1 contents were reduced, resulting in lower blood pressure.

5.3. *In Vivo* Antihypertensive Activity of SPPH Combined with Captopril

The antihypertensive effect of captopril is superior to that of SPPH with different doses in both short- and long-term animal tests (Wang, 2011). However, it is well known that chemically synthesized antihypertensive drugs have obvious side effects, and the use of health foods to regulate physiological conditions has become increasingly accepted by consumers in recent years. The ACE inhibitory peptides from food protein would be more widely recognized and consumed daily by its wide source, safe and reliable, and no toxic and side effects. Therefore, it is necessary to study the effect of the combination of SPPH and captopril on the blood pressure of SHRs, thus to reduce the harm to the body caused by long-term use of antihypertensive drugs.

5.3.1. Group of Test Animals

Ten-week-old female SHRs (165 ± 10 g) were used. All SHRs were kept in single cages and were allowed free access to the diets and water. Animals were kept under a 12 h light/dark cycle, 25 ± 1°C ambient temperature and 60 ± 5% relative humidity for 7 days prior to experimental use. SHRs were randomly divided into 3 groups with 3 rats per group: blank control (Group 1, same volume of distilled water), positive control (Group 2, Captopril 5 mg/kg body weight), SPPH + Captopril (Group 3, SPPH 1500 mg/kg body weight + Captopril 5 mg/kg body weight).

5.3.2. Effect of Single Administration of SPPH Combined with Captopril on SBP and Heart Rate of SHR

Compared with blank control, positive control (Group 2) and SPPH+Captopril (Group 3) presented antihypertensive activity within 3 to 6 h. The maximum decrease of SBP in SHRs with Captopril and SPPH+Captopril occurred after 3 h administration, which were 9.55 and 15.08 mmHg respectively. It was obvious that antihypertensive activity of SPPH+Captopril was better than that of Captopril. When time up to 6 h, SBP of SHRs with Captopril and SPPH+Captopril increased, but still significantly lower than that of blank control. After oral administration of Captopril and SPPH+Captopril, the heart rate of SHRs in each group was not significantly different from the blank control group within 6 h, thus indicating that both Captopril and SPPH+Captopril had no adverse effect on the heart rate of SHRs.

5.3.3. Effects of Long-Term Administration of SPPH Combined with Captopril for 24 Days on SBP and Heart Rate in SHR

Compared with the blank control, SBP of SHRs with Captopril and SPPH+Captopril after long-term administration for 6 days significantly decreased 17.31 and 20.03 mmHg, respectively. With the administration time prolonged for 12 days, SBP of SHRs with Captopril and SPPH+Captopril significantly decreased 19.98 and 25.90 mmHg, respectively. With the administration time prolonged for 24 days, SBP of SHRs with Captopril showed no significant difference with that for 12 days, whereas SBP of SHRs with SPPH+Captopril significantly decreased 28.87 mmHg. Thus, the combination of SPPH and Captopril had a synergistic effect on the reduction of SBP of SHRs, of which the antihypertensive activity was better than that of captopril alone. After oral administration of Captopril and SPPH+Captopril, the heart rate of SHRs in each group was not significantly different from the blank control group within 24 days, thus indicating that Captopril and SPPH+Captopril had no adverse effect on the heart rate of SHRs.

5.3.4. Effects of Long-Term Administration of SPPH Combined with Captopril for 24 Days on Plasma Lipid Level of SHR

After administration for 24 days, the TC, TG, LDL-C, and HDL-C in blank control group were 1.13, 0.42, 1.11 and 0.17 mmol/l, respectively. For Captopril group, the TC, TG, HDL-C and LDL-C of SHRs were 1.04, 0.42, 1.02 and 0.16 mmol/l, whereas the TC, TG, HDL-C and LDL-C of SHRs with SPPH + Captopril were 1.11, 0.42, 1.08 and 0.15 mmol/l respectively, showing no significant difference with those of blank control.

5.3.5. Effects of Long-Term Administration of SPPH Combined with Captopril for 24 Days on Endothelin-1 (ET-1) Content of SHR

Compared with the blank control, the Captopril and SPPH + Captopril groups showed lower ET-1 content. The contents of ET-1 in SHRs with Captopril and SPPH + Captopril were 19.12 and 16.64 pg/ml respectively. The results suggested that SPPH+Captopril showed better effect on the serum ET-1 content than Captopril alone, which might be the reason that SPPH+Captopril presented better antihypertensive activity.

5.3.6. Effects of Long-Term Administration of SPPH Combined with Captopril for 24 Days on Angiotensin II (Ang II) Content of SHR

In the classical renin-angiotensin system (RAS) pathway, ACE decomposes Ang I to produce Ang II. Ang II is the main active mediator, which exerts important biological effects on the body by binding to specific Ang receptors on the cell membrane *in vivo*. Ang II is an octapeptide with a strong contraction of blood vessels and pro-angiogenesis, which can lead to atherosclerosis. At the same time, Ang II can promote the secretion of some hormones in the body and indirectly increase the blood pressure; Ang II acts on the heart to strengthen the contraction of the heart muscle, causing ventricular hypertrophy and remodeling. Compared with the blank control, the Captopril and SPPH+Captopril groups showed lower Ang II content. The contents of Ang II for SHRs with Captopril and SPPH+Captopril were 18.60 and 14.91 pg/ml respectively. The results suggested that Captopril and SPPH + Captopril could reduce SBP by reducing ANG II expression, and SPPH + Captopril showed better effect.

In conclusion, the combination of SPPH and captopril had a synergistic effect on the reduction of SBP of SHRs for both short- and long-term administrations, and its effect was better than that of captopril alone.

REFERENCES

Ambrosi, V., Polenta, G., Gonzalez, C., Ferrari, G. & Maresca, P. (2016). High hydrostatic pressure assisted enzymatic hydrolysis of whey proteins. *Innovative Food Science and Emerging Technologies*, *38*, 294-301.

Arogundade, L. A., Mu, T. H. & Añón, M. C. (2012). Heat-induced gelation properties of isoelectric and ultrafiltered sweet potato protein isolate and their gel microstructure. *Food Research International*, *49*, 216-225.

Bonomi, F., Fiocchi, A., Frøkiær, H., Gaiaschi, A., Iametti, S., Poiesi, C., Rasmussen, P., Restani, P. & Rovere, P. (2003). Reduction of immunoreactivity of bovine β-lactoglobulin upon combined physical and proteolytic treatment. *Journal of Dairy Research*, *70*, 51-59.

Cui, S. S. (2014). The study on emulsifying properties of sweet potato hydrolysates with high pressure treatment, Xinjiang Agricultrual University. (In Chinese).

Dayem, A. A., Choi, H. Y., Kim, J. H. & Cho, S. G. (2010). Role of oxidative stress in stem, cancer, and cancer stem cells. *Cancers*, *2*, 859-884.

Doucet, D., Otter, D. E., Gauthier, S. F. & Foegeding, E. A. (2003). Enzyme-induced gelation of extensively hydrolyzed whey proteins by Alcalase:

peptide identification and determination of enzyme specificity. *Journal of Agricultural and Food Chemistry*, *51*, 6300-6308.

FAOSTAT. (2017). *Production of Crops*. http://faostat3.fao.org/browse/Q/QC/Ex.

Hernández-Ledesma, B., Davalos, A., Bartolome, B. & Amigo, L. (2005). Preparation of antioxidant enzymatic hydrolysates from α-lactalbumin and β-lactoglobulin. Identification of active peptides by HPLC-MS/MS. *Journal of Agricultural and Food Chemistry*, *53*, 588-593.

Hou, W. C., Han, C. H., Chen, H. J., Wen, C. L. & Lin, Y. H. (2005). Storage proteins of two cultivars of sweet potato (*Ipomoea batatas* L.) and their protease hydrolysates exhibited antioxidant activity *in vitro*. *Plant Science*, *168*, 449-456.

Ishiguro, K., Kurata, R., Shimada, Y., Sameshima, Y. & Kume, T. (2016). Effects of a sweetpotato protein digest on lipid metabolism in mice administered a high-fat diet. *Heliyon*, *2*, e00201.

Ishiguro, K., Sameshima, Y., Kume, T., Ikeda, K.-i., Matsumoto, J. & Yoshimoto, M. (2012). Hypotensive effect of a sweetpotato protein digest in spontaneously hypertensive rats and purification of angiotensin I-converting enzyme inhibitory peptides. *Food Chemistry*, *131*, 774-779.

Jung, W. K., Mendis, E., Je, J. Y., Park, P. J., Son, B. W., Kim, H. C., Choi, Y. K. & Kim, S. K. (2006). Angiotensin I-converting enzyme inhibitory peptide from yellowfin sole (Limanda aspera) frame protein and its antihypertensive effect in spontaneously hypertensive rats. *Food Chemistry*, *94*, 26-32.

Leonard, S. S., Keil, D., Mehlman, T., Proper, S., Shi, X. & Harris, G. K. (2006). Essiac Tea: Scavenging of reactive oxygen species and effects on DNA damage. *Journal of Ethnopharmacology*, *103*, 288-296.

Lorenzo, J. M., Munekata, P. E., Gómez, B., Barba, F. J., Mora, L., Pérez-Santaescolástica, C. & Toldrá, F. (2018). Bioactive peptides as natural antioxidants in food products–A review. *Trends in Food Science and Technology*, *79*, 136-147.

Mu, T. H., Liu, Y., Zhang, M. & Sun, H. N. (2014). Protein recovery from sweet potato starch wastewater by foam separation. *Separation Science and Technology*, *49*, 2255-2260.

Mu, T. H., Chen, J. W. & Mei, X. (2011). A method for extracting protein from sweet potato starch processing waste water, (Ed.) *N.I.P. Administration*, Vol. ZL201110190167.5. China. (In Chinese).

Mu, T. H., Tan, S. S. & Xue, Y. L. (2009). The amino acid composition, solubility and emulsifying properties of sweet potato protein. *Food Chemistry*, *112*, 1002-1005.

Norton, T. & Sun, D. W. (2008). Recent advances in the use of high pressure as an effective processing technique in the food industry. *Food and Bioprocess Technology*, *1*(1), 2-34.

Pedroche, J., Yust, M. M., Girón-Calle, J., Alaiz, M., Millán, F. & Vioque, J. (2002). Utilisation of chickpea protein isolates for production of peptides with angiotensin I-converting enzyme (ACE)-inhibitory activity. *Journal of the Science of Food and Agriculture*, *82*, 960-965.

Penas, E., Restani, P., Ballabio, C., Prestamo, G., Fiocchi, A. & Gomez, R. (2006). Evaluation of the residual antigenicity of dairy whey hydrolysates obtained by combination of enzymatic hydrolysis and high-pressure treatment. *Journal of Food Protection*, *69*, 1707-1712.

Ruan, C. Q., Chi, Y. J. & Zhang, R. D. (2010). Kinetics of hydrolysis of egg white protein by pepsin. *Czech Journal of Food Science*, *28*, 355-363.

Ruiz, J. Á. G., Ramos, M. & Recio, I. (2004). Angiotensin converting enzyme-inhibitory activity of peptides isolated from Manchego cheese. Stability under simulated gastrointestinal digestion. *International Dairy Journal*, *14*, 1075-1080.

Saito, Y., Wanezaki, K., Kawato, A. & Imayasu, S. (1994). Structure and activity of angiotensin I converting enzyme inhibitory peptides from sake and sake lees. *Bioscience, Biotechnology and Biochemistry*, *58*, 1767-1771.

Sarsour, E. H., Kumar, M. G., Chaudhuri, L., Kalen, A. L. & Goswami, P. C. (2009). Redox control of the cell cycle in health and disease. *Antioxidants and Redox Signaling*, *11*, 2985-3011.

Shackelford, R. E., Kaufmann, W. K. & Paules, R. S. (2009). Cell cycle control, checkpoint mechanisms, and genotoxic stress. *Environmental Health Perspectives*, *107*, 5-24.

Smith, D. G., Magwere, T. & Burchill, S. A. (2011). Oxidative stress and therapeutic opportunities: focus on the Ewing's sarcoma family of tumors. *Expert Review of Anticancer Therapy*, *11*, 229-249.

Smith, M. A., Harris, P. L. R., Sayre, L. M. & Perry, G. (1997). Iron accumulation in Alzheimer disease is a source of redox-generated from radicals. *Proceedings of the National Academy of Sciences of the United States of America*, *94*, 9866-9868.

Suetsuna, K. & Nakano, T. (2000). Identification of an antihypertensive peptide from peptic digest of wakame (*Undaria pinnatifida*). *The Journal of Nutritional Biochemistry*, *11*, 450-454.

Toyokuni, S. (1996). Iron-induced carcinogenesis: The role of redox regulation. *Free Radical Biology and Medicine*, *20*, 553-566.

Tsai, J. S., Chen, T. J., Pan, B. S., Gong, S. D. & Chung, M. Y. (2008). Antihypertensive effect of bioactive peptides produced by protease-facilitated lactic acid fermentation of milk. *Food Chemistry*, *106*, 552-558.

Valdez-Flores, M., Germán-Báez, L. J., Gutiérrez-Dorado, R., Medina-Godoy, S., Norzagaray-Valenzuela, C., Hernández-Verdugo, S., Reyes-Moreno, C. & Valdez-Ortiz, A. (2016). Improving bioactivities of Jatropha curcas protein hydrolysates by optimizing with response surface methodology the extrusion cooking process. *Industrial Crops and Products*, *85*, 353-360.

Wang, S. (2011*). The Study on Preparation of ACE Inhibitory Peptides from Sweet Potato and Its Antihypertensive Activity*, Chinese Academy of Agricultural Sciences. (In Chinese).

Wang, S., Mu, T. H. & Li, P. G. (2011). Preparation of Angiotensin I-converting enzyme inhibitory petide from sweet potato protein by optimizaiton of pepsin hydrolyzation. *Food Science and Technology*, *36*, 2-7.

Xiong, Z. D., Li, P. G. & Mu, T. H. (2009). The differentiation-and proliferation-inhibitory effects of sporamin from sweet potato in 3T3-L1 preadipocytes. *Agricultural Sciences in China*, *8*, 671-677.

Zhang, M. & Mu, T. H. (2017). Identification and characterization of antioxidant peptides from sweet potato protein hydrolysates by Alcalase under high hydrostatic pressure. *Innovative Food Science and Emerging Technologies*, *43*, 92-101.

Zhang, M. & Mu, T. H. (2016). Optimisation of antioxidant hydrolysate production from sweet potato protein and effect of *in vitro* gastrointestinal digestion. *International Journal of Food Science and Technology*, *51*, 1844-1850.

Zhang, M. & Mu, T. H. (2019). Improvement of thermal, microwave and ultrasonication pretreatment on the production of antioxidant peptides from sweet potato protein via *in vitro* gastrointestinal digestion. *International Journal of Food Science and Technology*, accepted.

Zhang, M., Mu, T. H. & Sun, M. J. (2014). Purification and identification of antioxidant peptides from sweet potato protein hydrolysates by alcalase. *Journal of Functional Foods*, *7*, 191-200.

Zhang, M., Mu, T. H. & Sun, M. J. (2012a). Sweet potato protein hydrolysates: antioxidant activity and protective effects on oxidative DNA damage. *International Journal of Food Science and Technology*, *47*, 2304-2310.

Zhang, M., Mu, T. H., Wang, Y. B. & Sun, M. J. (2012b). Evaluation of free radical-scavenging activities of sweet potato protein and its hydrolysates as affected by single and combination of enzyme systems. *International Journal of Food Science and Technology*, *47*, 696-702.

In: Proteases: Functions, Mechanisms and Uses ISBN: 978-1-53615-854-0
Editor: Nelson Pérez Guerra

Chapter 7

NATURAL PROTEASE INHIBITORS AND SULPHUR-CONTAINING AMINO ACIDS

Mammo Mengesha Erdaw*
Ethiopian Institute of Agricultural Research,
Debre-Zeit Agricultural Research Centre, Debre-Ziet/Bishoftu, Ethiopia

ABSTRACT

Some of the constituents of food or mixed-feed have not yet been properly utilized because of the presences of anti-nutritional factors (ANF). One of the ANF that are commonly and varyingly distributed throughout the nature is called protease inhibitors (trypsin inhibitor and chymotrypsin inhibitor). Trypsin inhibitor (TI) is almost negligible in cereals, but highly concentrated in legume grains, particularly in soybeans. The TI is mainly located in the protein bodies of the cotyledon. The ANF can generally be grouped as heat-labile, heat-resistant or heat-tolerant. Depending on the processing conditions, certain amounts of TI is always available in the consumable products, for example, a soybean-protein isolate contain up to 40% of the TI activity. Bowman-birk inhibitor (BBI), a rich in the contents of cysteine has more resistance to heat, compared to Kunitz trypsin inhibitor (KTI). Methionine, cysteine, homo-cysteine, and taurine are the

* Corresponding Author's E-mail: mammomerdaw@gmail.com or leulmammo@yahoo.com.

four common sulphur-containing amino acids (AA), but only the first two are naturally incorporated into proteins. A protein from soybean is not an ideal because of inadequacy of methionine (Met), but the BBI and to some extent KTI are positively contributing the quality due to their higher contents of cysteine. Digestibility of methionine could not be affected by the availability of TI, in the raw soybeans. The TI is generally responsible to negatively affecting endogenous trypsin enzymes and thereby creates pancreatic hypertrophy of some animals. However, there is uncertainty with regards to the biological threshold levels of this TI, in foods/diets of susceptible animals. Unlike cysteine the sulphur of methionine is not highly nucleophilic, and therefore, methionine from unheated soybean would slowly be liberated and absorbed. Although methionine is always an initiator to synthesis that of all types of eukaryotic proteins, cysteine has ability to form disulfide-bonds and also plays a crucial role of protein-structuring. It is therefore concluded that despite the fact that their availability certainly enriches the sulphur-containing amino acids, protease inhibitors are generally and usually impairing the nutritional values of most legume grains.

Keywords: anti-nutritional factors, heat-treatments, protease-inhibitors, sulphur-containing amino acids

1. Introduction

Any feed, food or ingredient that is being used as a plant-source protein, particularly for the non-ruminant animals is mainly derived either from leguminous grains or oilseeds. However, one of the limitations to an increased use of grain legumes is the presence of anti-nutritional factors (ANF), which reduce the nutritional quality, even sometimes after processing. Anti-nutritional factors, such as trypsin inhibitors (TI), phytic-acid and cyanogen are some of important ones. These ANF draw their negative effects via different mechanisms, for example through binding the digestive enzymes and nutrients or increasing the gut viscosity (Ao, 2011).

Protease inhibitors (trypsin and chymotrypsin) are the ones among the other ANF that are varyingly found in legume grains. Although it has been suggested that proteinaceous anti-nutrients could be inactivated by heating, heat treatment generally may not be considered as an effective approach to completely inactivate all types of ANF. Clemente et al. (2008) reported that Bowman-Birk inhibitors (BBI), which are a group of protease inhibitors, always exhibit a considerable resistance to heat treatments.

Exogenous/microbial protease enzyme is one of the newest technologies, with animal or vegetable proteins as its substrate. These can break down the ANF associated with various proteins (Roux, 2016). Microbial protease improves the digestion of proteins and increase amino acid availability, which also helps to release nutrients (Roux, 2016).

Protease enzymes basically catalyzes the process of hydrolysing the peptide and protein bonds and represents one of the most important groups of commercial and industrial enzymes (Sumantha et al., 2006). Raw ingredients with low amino acid digestibility could greatly respond to an exogenous protease. Microbial proteases are basically helping the producers to manage the nutritional risks that are associated with the quality of feedstuff and also allow them to utilize all available feed ingredients (Roux, 2016).

2. Anti-Nutritional Factors

Some of the legume grains to be used by humans or non-ruminant animals are soybean, peas, faba-bean, lupin, common-vetch and grass-pea. However, these grains contain a wide variety of ANF, such as oligosaccharides, proteinaceous compounds, lectins, amylase inhibitors, and phytic acid (Ramadoss and Shunmugam, 2014). Among which protease inhibitors are widely distributed throughout nature and are found in plants, animals and microorganisms (Burns, 1987).

Anti-nutritional factors are chemical compounds in plants and are biologically active. The ANF or anti-nutrients are substances of which directly or through their metabolic products, interfere with the nutrient assimilation. Plants commonly synthesize a range of biologically active metabolites as part of their protection against any attacks by the herbivores, insects and pathogens or as a means to survive in an adverse growing conditions (Khokhar and Apenten, 2003; Soetan and Oyewol, 2009). The levels or concentrations of these anti-nutrients are usually varying with the species of plants, cultivars or post-harvest treatments (Akande et al., 2010).

Francis et al. (2001) reported that use of plant-derived materials, such as legume seeds, oilseed cakes, leaf meals, leaf-protein concentrates, and root-tuber meals, as ingredients are limited by the presence of a wide variety of anti-nutritional substances. These all ANF may be majorly categorized as heat-stable, heat-tolerant or heat-resistance (Table 3). For example, in the soybean grain there are three well-known ANF, including TI, lectins and phytate (Chen et al., 2013; Pettersson and Pontoppidan, 2013). Among which the most important ones, in soybean meal is TI (Newkirk, 2010; Dourado et al., 2011). All types of TI are not completely inactivated by the heat treatments (Clemente et al., 2008). Phytic acids and lectins are the other ANF that are commonly and varyingly found in most legume grains, for example their concentrations in raw soybean grains were 6 ± 8 to 17 ± 5 g/kg and 10 to 20 g/kg, respectively (Clarke and Wiseman, 2000).

Furthermore, soybean seed also contains a high amounts of oligosaccharides and are poorly digested and thereby causes for the low utilization of energy when fed to the non-ruminant animals (Lesake et al., 1995). Therefore, to enhance the nutritional and health-promoting aspects of soybean proteins, better understanding is initially needed about the factors that are negatively contributing to the reduced nutritional quality of the ingredient (Friedman and Brandon, 2001). Despite the efficiency of thermal treatments, in reducing the protease inhibitors, residual inhibitors (10 - 20%) are still existed in the processed-products (de-Carvalho et al., 1998). But, these residues can be further reduced by supplementing the microbial enzymes (Neus et al., 2005; Erdaw et al., 2017a; Erdaw et al., 2018).

3. Protease Inhibitors

The protease inhibitors are protein-based molecules and are found in plants (Tacon, 1997). Generally protease inhibitors that mainly found in such plants contain little or no carbohydrates and have a molecular weight ranging from 4 to 80 kDa. All protease inhibitors have one or more peptide bonds (reactive-site), which interacts with the corresponding active-sites of the enzymes. The reactive-sites are normally found in a close association with half cysteine residues that linked by an S-S bonding bridges (Laskowski and Kato, 1980). Often two or more enzymes can be inhibited by the same protease inhibitors. In this instance, different enzymes can compete for a substrate.

A protease inhibitor is biochemically categorized into two, such as KTI that mainly inhibits the trypsins. Secondly, BBI, with the cysteine residues are also have capability to inhibit both chymotrypsins and trypsins at independent binding sites. Binding such sites by the ANF means the normal digestion could be negatively affected. In contrast, Tan-Wilson and Wilson (1986) suggested that the BBI, and to some extent the KTI, are always contributing to the nutritional quality of the feed ingredients, including soybeans by the virtue of their high cysteine contents, which contains sulphur in it.

4. Trypsin and Chymotrypsin Inhibitors

Natural trypsin inhibitors (TI) also known as serine protease inhibitors (serpins) are the largest and most diverse family of protease inhibitors. This is a KTI group and mainly inhibits the trypsin enzymes. These can be found from the four main sources, such as bovine pancreas, ovomucoid, soybean, and lima bean. Each of this acts as a competitive substrate in binding with its serine protease to form an inactive complex (Zhou et al., 1989). The TI from soybean is a monomeric protein containing 181 AA residues in a single polypeptide chain that are cross-linked by the two disulfide bridges. These reduce the biological activities of endogenous enzymes, particularly the endogenous trypsin enzymes and then by doing these actions, these are considered as ANF.

A trypsin is a protein-digesting enzyme present in pancreatic juices that is being secreted into the small intestine. Initially, pancreas secretes the trypsin as an inactive pro-enzyme, called trypsinogen. However, once in the intestine it is usually changed into active trypsin enzymes, and in turn to break down the proteins into smaller and absorbable forms. As shown in Table 2, TI is varyingly concentrated in legume grains, which is ranging between 3 - 8 TIU/mg (lentil) and 43 - 84 TIU/mg (soybean). The concentration of TI is also naturally varying based on the processing conditions of feed/food ingredients (Table 1).

Table 1. Trypsin inhibitor at very common products of soybean

Products of soybean	Trypsin inhibitor activity	
	mg/g sample	mg/g protein
Whole soybean	16.7 - 48.2	34.7 - 122.6
Raw soybean flour	52.1	104.26
Toasted soybean flour	7.9-9.4	15.9

Sources: As summarized by Gilani et al. (2005).

Table 2. Contents of trypsin inhibition in the grains of different species

Ingredients, grains	TI (TIU/mg sample)
Soybean	43 - 84
Common beans	21 - 25
Lathyrus cultivars	19 - 30
Chick peas	15 -19
Peas	6 - 1 5
Lentil	3 - 8
Faba been	5 - 10

Source: Guillamon et al., 2008.

Although the recent evidence clearly indicates that TI, from raw grain of soybeans is the major factor that is responsible for the pancreatic hypertrophy of susceptible animal followed by growth inhibitions, there is still uncertainty regarding the biological threshold levels of TI in diets, at which these biological effects could occur on the different age groups and species of the animals (Rackis, 1981).

Previously when only a small proportion of the diet consisted of soybean products, the fluctuations due to TI were largely undetectable in terms of altered performance of birds (Clarke and Wiseman, 2005). However, inclusion rate of soybean products in diets of animals including human is recently increasing, which creates doubts on the threshold levels of TI concentration.

The predominant TI in leguminous grains, including soybeans and derived materials are basically proteins, and are mainly located in the protein bodies of the cotyledon (Anderson and Wolf, 1995). Hence the TI certainly tends to fractionate at times of both storage and processing. Protease inhibitors make up 0.2 - 10.0% of the total seed protein of edible dry beans of various species. Concentration of TI, in soybeans has been reported to vary from 17 - 48 mg/g of the sample or from 37 - 123 mg/g of the protein. When soybeans are processed into raw defatted flour, none of the TI is removed; instead it becomes concentrated to the levels of 28 - 52 mg/g of the flour, which corresponds to levels of 58 - 104 mg/g of the protein (Gilani et al., 2005). The extent to which TI activity is reduced by heat depends on its initial level that presents in the starting material, temperature, heating time, particle size, moisture and, probably the crop species and cultivars (Gilani et al., 2005).

Trypsin and chymotrypsin are both serine protease with a high sequencing and structural similarities, but the main difference is their substrate specificity (Ma et al., 2005). The same authors added that trypsin and chymotrypsin have distinct dynamic signals in the two loop regions, which are in turn highly correlated with motions of certain residues in the binding pockets.

5. Natural Protease Inhibitors and Sulphur-Containing Amino Acids

Residues of methionine, cysteine, or tryptophan that are close to or within the active sites of enzymes are normally involved in the oxidative inactivation of the enzymes. Under the non-oxidative conditions, sulphur-containing amino acids, such as methionine and cysteine exhibited the most specific activities towards their substrates. These actually allowed by the small amino acids and by amino acids with amides and aliphatic side chains

(Li et al., 2013). Sulphur-containing amino acids are also called sulphur amino acids which contain sulfhydryl, and these plays the crucial roles in process of protein structuring, metabolising, immunising, and oxidizations (Bin et al., 2017).

Before many years ago, Frost and Mann (1966) reported that the TI diminishes the conversion of methionine to cysteine through using cystathionine pathways, which is either by interfering through incorporating the cysteine into protein or by interfering directly into a cystathionine synthase. Methionine is usually converted into homocysteine, which in turn donates a sulphur group to the serine amino acids that ultimately form the cysteine (Teeter et al., 1978).

Soybean with higher protein contents had a significantly lower percentage of sulphur-containing amino acids, while others with lower protein exhibited a higher content of cysteine and methionine (Table 3).

Nitrogen application logically elevated the protein contents of the crop, but concentration of sulphur-containing amino acids, including the BBI, which is rich in cysteine became lower (Krishnan et al., 2005). Before years ago, Griffiths (1984) also reported that contents of both trypsin and chymotrypsin inhibitors appeared to be independent of the crude protein contents. These protease inhibitors were also largely unaffected by dry heat, but were rapidly denatured by autoclaving. Similarly, Kochhar et al. (1988) studied on varieties of cowpea and found that there was no relationship between contents of protein and the methionine.

Except methionine, ileal digestibility of a total AA was lowered for broilers when they fed diets containing the raw soybean (Erdaw et al., 2017a; Erdaw et al., 2018), as shown in Table 5. Furthermore, Kwong and Barnes (1963) previously reported that feeding the rats with unheated soybeans didn't selectively impair the availability or tissue utilization of methionine, but cysteine was blocked from the metabolic utilization of the protein synthesis.

Cysteine was increasingly accumulated in the small intestine with increased levels of dietary inhibitors (Krogdahl et al., 1994). Previously, Broadway and Duffey (1986) found an improved performance of rats fed diets, containing TI from soybean and supplemented with methionine and they concluded that availability of sulphur-containing amino acids are important to synthesis endogenous enzymes. Except cysteine, the digestibility of the total AA was similarly affected in fish fed diets containing TI in soybeans (Krogdahl and Berg-Lea, 1992). Approximately 76% of the KTI and only 4.8% of the BBI were survived passing through the small bowel of orally given rats (Hajos et al., 1995). These all also provide the support to the hypothesis that the growth depression caused by the TI was the consequence of an endogenous loss of AA in the form of enzymes that are being secreted by a hyperactive pancreas.

The pancreatic enzymes, such as trypsin and chymotrypsin, are particularly rich in sulphur-containing AA. Therefore, TI stimulates for further secretion of digestive enzymes from the pancreas by a negative feedback mechanism, which is mediated by cholecystokinin. The effects of a hyperactive pancreas would be to divert these AA from the synthesis of body-tissue protein to the synthesis of endogenous enzymes, which are subsequently lost in the feces (Green and Lyman, 1972; Friedman and Brandon, 2001).

Unless legumes grains are heat treated before consumption, methionine would be slowly liberated by the proteolytic enzymes and its absorption is also being delayed (Oke, 2007). One of the effects of the inactivation of digestive endogenous enzymes in the intestine by the TI is the stimulation of trypsin and chymotrypsin secretion from the pancreas. These can create increased demand for the sulphur-containing amino acids. In turn, this leads to increase the loss of both endogenous nitrogen and sulphur elements. Finally, the inhibitors can stimulate the release of cholecystokinin into the blood stream, which further increases pancreatic secretion (Green and Lyman, 1972; Hill, 2003).

Protease inhibitors biochemically be grouped into 2, such as KTI (with a molecular weight of 20-25 kDa) that inhibits only trypsin and the second one is BBI, which is rich in cysteine residues (with a molecular weight of 6-10 kDa), is capable of inhibiting the chymotrypsin as well as trypsin at independent binding sites (Makkar et al., 2007). The KTI of soybean has several polymorphic variants, such as Tia and Tib, which differs by 9 AA. These are the two main types and have different proteomic profiles and TI activities (Lee et al., 2012). Weight of pancreas from rats, which fed diets containing unheated soybean supplemented with sulphur-containing AA was higher than those of rats fed un-supplemented diets (Gumbmann and Friedman, 1987).

Soybean protein is not an ideal protein because it is deficient in methionine (Friedman and Brandon, 2001). However, the BBI, and to some extent the KTI contributes to the nutritional quality of soybeans (Tables 6 and 7) by the virtue of their high cysteine contents which could supplements the low amounts of sulphur-containing AA in the storage proteins that comprise the bulk of the protein reserve in the seed (Tan-Wilson and Wilson, 1986). Addition of methionine to soybean milk formula increased the nitrogen retention, but did not have the same effect when added to the isolate soybean proteins (Jordão and Marchin, 1998). With regards to this, Kanjanapruthipong (1998) tested by supplementing the milk with a replacer containing soybean protein and sulphur-containing amino acids considerably improved the protein quality of the milk replacer for the pre-ruminant calves. Absorption of Zinc was also enhanced in diets of rats supplemented with methionine or cysteine (House et al., 1996).

6. Methionine and Cysteine

Methionine, cysteine, homocysteine, and taurine are the 4 common sulphur-containing AA, but only the first 2 are naturally incorporated into proteins. Methionine is a crucial amino acid for most of the animals and it cannot be synthesized, with sufficient amounts that could not maintain both normal growth and production, but cysteine is a semi-essential amino acid because it can be synthesized/produced. Within proteins, many of the methionine residues are buried in the hydrophobic core, but the exposed ones are susceptible to oxidative damaging. Generally, methionine and cysteine are anti-oxidative, and also alleviate the oxidant stress that could induced by various oxidants and protect the tissue from the damage (Bin et al., 2017).

Although it will react with some electrophilic centres, unlike cysteine the sulphur of methionine is not highly nucleophilic. Cysteine and methionine have remarkably similar functions and both of them are important antioxidants that contribute to the structure and stability of proteins. Their relative ease of reversible oxidations and reduction fills both of these amino acids with the ability to serve as regulatory switches and signals in increasingly broad venues (Kim et al., 2014). However, cysteine is a well-recognized one for the ease of its oxidation.

It is not usually appreciated that methionine can readily be oxidized to methionine (Vogt, 1995). Cysteine, by virtue of its ability to form disulfide bonds, plays a crucial role in structuring proteins and folding pathways (Levine et al., 1996; Brosnan and Brosnan, 2006). It is well established fact that methionine is the initiating AA in the synthesis of every eukaryotic protein (Sasaki and Nakashima, 2000). However, both deficiency and excess of methionine and cysteine in diets affects the normal growth of animals (Bin et al., 2017). Animal-based foods are generally good sources of both methionine and cysteine. These contain the element sulphur, which plays important roles in determining the shapes of new proteins (Marie, 2017).

7. Heat Treatments and Residuals of Some Anti-Nutritional Factors

Anti-nutrients can have negative impacts on nutritional status (Roos et al., 2013). The contents of anti-nutrients, including in the processed foods or feeds depends on the ingredients and processing conditions. Dry heating of the seeds and the meal was not completely effective in inactivating both the TI and chymotrypsin inhibitory activities (Mulimani and Paramjyothi, 1994). Most commercially heated meals retain at least up to 20% of the protease inhibitors. Previously, Rackis et al. (1986) suggested also that all commercially processed soybean products retain some TI in it. Additionally, Clarke and Wiseman (2005) analyzed samples of processed soybean meals and actually found a varying concentrations of TI (1.1 to 3.6 mg/g) and confirmed the same trend that a pancreas to body weight ratio was also increased.

Table 3. Effects of levelled extrusions on soybean products

Treatments	Extrusion (°C)	TI (mg/g)	UA, ΔpH	BWG	FI/bird	FCR
1 (Raw-soy)	-	37.92	2.03	1502 ± 32	3815	2.53
2	118	9.41	1.08	1890 ± 40.9	3627	1.91
3	120	6.68	0.85	1897 ± 36.5	3565	1.87
4	122	1.66	0.10	2056 ± 28.6	3534	1.71
5	126	1.26	0.09	2068 ± 30.7	3594	1.73
6	140	N.D.	0.05	1988 ± 37.1	3403	1.71
Parameters		**Raw-Soybean**	**SBM**	**Enzyme-treated-SPC**	**Alcohol-extracted**	**SPI**
Crude protein, %		33 - 37	42	55 - 60	63 - 67	> 85
Crude Fat, %		17 - 20	0.9	2.5	0.5 - 0.3	0.1 -1.5
Oligosaccharides, %		14		< 1.0	< 3.5	< 0.4
Trypsin inhibitors, mg/g		25 - 50	1.6 - 5.0	1 - 2	2 - 3	< 1
Lectins, ppm		2100 - 3500	20 - 600	< 1.0	< 1.0	< 1.0
Saponins, %		0.5	0.6	0	0	0
Phytic acid bound, %		0.38	0.42	0.6	0.6	> 85

Source: Summarized from Van Eys et al. (2004). UA = urease activities; TI = trypsin inhibitors; FI = feed intake; BWG = body weight gains; FCR = feed conversion ratio; SPI = soy protein isolates; SPC = soy protein concentrate; ΔpH = change of the temperature; SBM = soybean meal.

Roos et al. (2013) evaluated 36 samples of processed-products and found that residuals of trypsin and chymotrypsin inhibitors and lectins were sufficiently enough to reduce the pancreatic function of young infants. The same authors conclude about the need of a threshold limits for the residual II in foods of infants. Additionally, Roos et al. (2013) reported that residue of phytates in the processed foods of both cereal and legume-based products was ranging between 68 and 1536 mg/100 g.

However, Eva et al. (2010) stated that due to the wet heat treatments, most of the TI became inactivated. Kelli et al. (2011) added also that the TI activity was totally inactivated by a pressure-cooking. Effects of heat treatment on TI however vary with that of both plant species and structural formula of each TI (Ezeagu, 2006). Depending on their modes of preparations, soybean isolates may contain the TI activity as high as 40% of that mostly found in the raw soybeans (Baker and Rackis, 1986). In contrast to this, Gilani et al. (2005) reported that the presence of residual contents of TI in isolate protein from heat-treated soybean.

Brandon et al. (1991) measured the residual contents of TI in processed products of soybean for infants and found that the infants consume about 10 mg of trypsin plus chymotrypsin inhibitors per day. These authors concluded that heating or fractionation of protein isolates does not completely inactivate or remove these inhibitors. So that residual amounts of inhibitors are always consumed by animals and humans. Soybean products planned for human consumption, such as soybean protein concentrate (70% CP), soybean protein isolates (90% CP), and soybean-based infant formulas have received sufficient heat treatment to denature up to 80% of the TI activity present in raw soybean flour.

Residuals of TI from soybean products combining with endogenous trypsin in the small intestine could form an inactive complex to reduce the digestibility of proteins (Van-Eys et al., 2004). For example, the TI could also correlate with the chickens' syndrome of "rapid feed passage" (Ruiz and Belalcazar, 2005). A combined use of two or more methods may be more efficient in inactivating the ANF, and Akande and Fabiyi (2010) suggested also that a single method may not be effective to removing the anti-nutritional substances. In contrast, Luo and Xie (2013) reported that some treatments even increased the level of ANF in green and white faba beans.

A method that usually used in determining the residual urease activity of soybean products, which is an indirect indicator to assess whether TI is also sufficiently present in soybeans could be destroyed by heat processing. Enzymes, urease and TI could be deactivated during heating. Burnes (1987) previously reviewed on the physiological significance of residual protease inhibitors from commercially processed plant proteins and human foods, and then conclusively reported that the concentration of protease-inhibitors activities was defectively low of which have no nutritional concern in animals or humans. Additionally, Flavin (1982) reviewed and reported that the effects of soybean TI, in soybean protein does not appear to be a potential hazard to human being and hence efforts to eliminate these inhibitors does not seem to be necessary for humans. Protein quality is very important in dictating the anti-nutritional effects of plant proteinase inhibitors (Broadway and Duffey, 1986).

8. Feeds or Foods with Trypsin Inhibitors and the Extents of Inactivation

As shown in Table 4, heat processing is widely used as an effective means of inactivating the thermo-labile ANF, particularly from the legume grains. Carvalho et al. (2013) reported also that anti-nutrients from hulled and hull-less soybean flours of genetically improved soybeans were inactivated at 150 °C for 30 minutes. However, Osborne and Mendel (1917) noted that dry heat was less effective than the cooking (moist heat) for the improvement of growth promoting action in soybeans. Soybean flour treated by moist heat (toasting) had destruction, at least 95% of the TI, whereas toasted soybean flour still had much of the TI (Rackis et al., 1975). Marsman et al. (1997) suggested that there is uncertainty with regards to the TI, in a heat-treated SBM, which are responsible for the variations that observed within the *in vivo* and *in vitro* protein digestibility. Furthermore, Khokhar and Apenten (2003) added that there is always occurrence of the TI compounds in both fresh and in the processed foodstuffs of human diets.

Some other ANF are partially and completely heat-stables and these are BBI, oligosaccharides and phytate (Clemente et al., 2008). Although Rackis and Gumbmann (1981) previously reported that the activity of TI, in soybeans can be readily be eliminated by ordinary cooking and moist-heat treatment, Clemente et al. (2008) suggested that BBI, a group of TI, are certainly be heat-stable. In this case, Krishnan et al. (2005) added that BBI is rich in cysteine and is better than the KTI in surviving the heat processing

(Tan-Wilson and Wilson, 1986). Livingstone et al. (2007) almost concluded that eliminating the BBI from soybeans has proven to be more difficult.

Table 4. Heat-labile and heat-stable anti-nutritional factors (ANF)

Ingredients	Heat-labile	Heat-stable
Soybean	Trypsin inhibitor, lectins, goitrogens; anti-vitamins	Saponins, NSP, phytate, phytoestrogens, protein antigens, tannins; estrogens
Rapeseed	Glucosinolates	Phytate, tannins, sinapine, fibre, phenolic compounds
Peas	Protease inhibitors, lectins cyanogens	Saponins, starch, phytic acid, protein antigens
Flax	Cyanogenic glycosides	Mucilage and NSP

Source: Murray (2009).

Some of the TI, glycinin antigens, oligosaccharides, and saponins, non protein AA, total phenolics, tannins, phytates, saponins, pyrimidine glycosides, alkaloids, cyanogenic glycosides, and flatus-producing oligosaccharides are heat-resisting ANF and could be remained in foods or feeds even after heat treatments (Liener, 1994). Heat treatment doesn't completely eliminate these factors and hence they may decrease the protein solubility. Despite the efficiency of thermal treatment to reducing the protease inhibitors, around 10 - 20% residual inhibition is still maintained after heating (de-Carvalho et al., 1998). Processing, based on the use of heat or fractionation of protein isolates, does not completely inactivate or remove these inhibitors, so that residual amounts of inhibitors are always consumed by animals and humans (Brandon et al., 1991).

Activities of TI varies with source of beans and processing conditions. Because of the wide variations of the processing conditions, the actual amount of TI ingested by poultry may vary considerably between the batches. Karr-Lilienthal et al. (2004) evaluated samples of SBM produced by Argentina, Brazil, China, India, and the United States for their TI contents through concentration of urease activity and then found a pH change of 0.2, which potentially showed a higher urease activity or indirectly a higher TI contents between the analyzed samples. As Gilani et al. (2012) reviewed and reported, there is no guarantee that each and every commercial product would be properly processed. The products would contain their own residual levels of TI, which may vary greatly with the extent of the heat and other processing conditions used in the preparation of soybean products. For example, several commercial soybean beverages have been reported to contain a significant level of residual TI activity (up to 71 and 64% of that of whole soybeans). Moreover, Vidal-Valverde et al. (1997) added that the TI was just partially removed from the seed of raw faba bean after dry-heating.

The TI activity value, which is usually below 2.5 mg/g of soybean, is considered to almost avoiding the syndrome on animals (Van-Eys et al., 2004). In relation to this, Clarke and Wiseman (2007) measured the digestibilities of methionine, cysteine, glutamine, lysine and valine when the TI concentration, in diets were 14.8, 9.6, 4.5 and 1.9 mg/g of full-fat soybean and they found that the highest digestibility was recorded when the TIA value was 1.9 mg/g.

9. Anti-Nutritional Factors and Their Inhibiting Mechanisms

Depending on the structures of an individual compound, which can be ranged from a high molecular-weight protein to the simpler AA or oligosaccharides, its biological effects could greatly vary (Khokhar and Apenten, 2003). One of the modes of actions of the enzyme inhibitors, for example TI is by making a strong attachment at the active sites of the endogenous enzymes.

Substrates are then prevented by the TI from being approached to the active sites, which could lead to effectively inactivate the endogenous enzymes. Anti-nutritional factors elicit their negative effects via different mechanisms, including the binding of the digestive enzymes and nutrients or increasing the gut viscosity (Ao, 2011). Approximately 80% of the TI is caused by KTI, which is strongly inhibiting the endogenous trypsin enzymes and thereby reduces the feed intake and other performances of animals through diminishing their digestion and absorption.

Table 5. Types and levels of heat treatments and inactivation of anti-nutritional factors

Ingredients	ANF	T (°C)	Duration (min)	Results	References
Soybean	TI	97	30	Insufficient to inactivate the TI	1*
	TI	Boiling water	30	Insufficient to inactivate the TI	1*
Soybean flour	TI	Autoclaved, 121°C	5	55% of IT inactivated	1*
Soybean flour	TI	toasted	-	10% of residual	6*
Soybean grains	TI	Autoclaved, 121°C	5	75% of IT inactivated	1*
Soybean	TI	Toasted	-	2.59 mg Tl/g	2*
Full-fat soybean	TI	Toasted	-	2.74- 3.28 (TIU/mg)	3a*
Full-fat soybean	UI	Toasted	-	0.02 - 0.30 (UI)	3a*
SBM	TI	©	-	2.45 to 1.98	4b*
	TI	©		TIA < 1.5 mg/g	5*
	TI	©	-	3.5 to 2.8	7*
Raw faba-bean-seeds	TI	Autoclaved, 120°C	15	Half of TI eradicated	8*
Grains	Phytates	150°C	60	Showed thermal decomposition	9*
Cereal bran	TI	Wet-heated, 110°C	25	reduced by 83.07%	10*

ANF = anti-nutritional factors; TI = trypsin inhibitors; UI = urease index; © = C° (Temp) is as per recommendation; SBM = soybean meal 1*) Carvalho and Sgarbieri, 1997); 2*) Sessa and Bietz, 1986; 3*) Ravindran et al., 2014a; 4*) Ravindran et al., 2014b; 5*) Varga-Visi et al., 2009; 6*) Anderson, 1986; 7*) García-Rebollar et al., 2016; 8*) Vidal-Valverde et al., 1997; 9*) Daneluti and Matos, 2013);10*) Kaur et al., 2012.

The other effects of KTI are induction of pancreatic enzymes (hyper secretion), a quicker stimulation of pancreas growth, hypertrophy and hyperplasia. Due to these, inclusion of high levels of raw soybean may not be advised to be used by the mono-gastric animals (Mikić et al., 2009). Phytates are also impeding the nutrient digestions through electrostatic mechanisms, which reduce the solubility of proteins. Consequently, this increases the loss of endogenous nutrients (Cowieson and Ravindran, 2007).

Table 6. Most suggested and recommended levels as well as indicators of a typical quality soybean product

Total lysine	**> 2.85% (dry matter basis)**
Digestible lysine	≥ 88% of total lysine
Ash	< 7.5%
Protein solubility in 0.2% KOH	78 - 85%
Urease activity	0.0 - 0.10 pH unit rise
Trypsin inhibitors	1.75 - 2.50 mg/g
Colour	Light to light-brown

Source: Summarized from Van Eys et al. (2004).

Scholars (Dale, 1996; Pettersson and Pontoppidan, 2013) reported that oilseed meals, particularly soybean meal (SBM) constitute as a major part, compared to other plant-protein source feeds, and is the most commonly used feeds of poultry and pigs. The SBM is a by-product that remained after extracting the oil. Common parameters and characteristics of a typical quality commercial soybean meal is shown in Table 6. However, as found in feeds, prepared from soybean-products, including the processed ones (commercial SBM) are still containing both heat-stable and residual of ANF. These might have potentials to reduce the nutrient utilizations. For example, piglets fed on the commercial SBM, as feed showed low tolerance and the conclusion was that soybean meals might contained a high level of several heat-stable ANF (Li et al., 1990). The procedures described in determining the concentration of trypsin inhibitors are based on the ability of the inhibitors to form a complex with the endogenous enzymes and thus to reduce the enzyme activities (Van-Eys et al., 2004).

10. Over- or under-Heating of the Feeds or Foods and the Contents of Anti-Nutritional Factors

Heat treatment is generally considered as one of the most effective approaches to inactivate some of the TI, but not the phytates, or Bowman-Birk inhibitors, oligosaccharides, and antigenic proteins. Furthermore, either under- or over-heating the feed ingredients while processing has frequently been reported to reduce the nutritional values of SBM for poultry (Perilla et al., 1997). The practice of using chemicals, such as solvents, in extracting the oil is questionable as the residual chemicals in the SBM may pose the health risks.

One of the strategies to improving such feed or food ingredients might be advised to use biotechnological techniques, such as genetic modification of the feedstuffs, or feed additives (enzymes). However, before intervening with the application of biotechnological techniques (enzymes), information regarding the physical and chemical properties of feed ingredient is required (McDonald et al., 2002).

Evidence is steadily emerging that the use of new-generation enzymes is one of the most promising approaches to enhance the productivity of non-ruminant animals through improving the nutritional values of feeds (Adeola and Cowieson, 2011). Microbial protease is potentially protein-digesting enzyme that is used in pig and poultry diets to break down the stored proteins and proteinaceous anti-nutrients in various plant materials (Barletta, 2011; Ao, 2011). In addition to releasing phosphorus, supplementation of microbial phytase could also improve the digestibility of CP and amino acids of plant-based proteins (Barletta, 2011; Guggenbuhl et al., 2012).

Additionally, variation was not observed in TI activities in a soaked seed as compared to the dried seeds of Mucuna pruriens, but TI values were varied due to germination. The TI activity decreased with an increase in germination time.

11. Practices of Plant Breeding to Reduce the Antinutritional Factors

Soybean cultivars for the reduced amounts of antinutritive factors resulted in the development of KTI-free soybean cultivars, which are suitable for thermal processing at lower temperatures and during a shorter period of time. Perić et al. (2011) reported that the TI content in the newly developed soybean cultivars was reduced by 50%, compared with the conventional cultivars. However, researchers suggested into such breeding for the low TI cultivars currently has limited application as TI contribute a major proportion of sulphur-containing AA, including methionine and cysteine (Clarke and Wiseman, 2000).

The TI negatively affects the functions of the pancreas, in terms of synthesis and secretion of digestive enzymes, including proteases. As an optional solution, some geneticists and crop breeding companies have developed seeds that have very low ANF. However, the TI are very rich in sulphur-containing amino acids as they may account for about 40% of the total cysteine contents of some legume proteins (Nabil, 2014), which could show that breeding may not be a strategic solution in solving the negative effects of ANF, particularly the TI in soybeans.

12. Natural Protease Inhibitors on Pancreas

Following the earlier discovery by Chernick et al. (1948), who found the reason why the chicks' pancreases were enlarged in sizes was due to the presence of TI in the raw soybeans, it was further understood that such enragements of the pancreas was the mechanism of enabling it to produce more enzymes as the compensation. When an endogenous trypsin is inhibited by TI, cholecystokinin production is automatically be enhanced, resulting in an increased production of the pancreatic digestive enzymes. However, Pettersson and Pontoppidan (2013) reported that the negative impacts of TI in the non-ruminant animal diets could be reduced by supplementing their diets with microbial protease.

As reviewed and reported by scholars (Liener and Kakade, 1980; Ge and Morgan, 1993), the consequences of TI on pancreases have been confirmed not only in chickens, but also in rats, mice and young pigs. Grant (1989) added that the dietary raw soybean or TI do alter the pancreatic secretion, which lead to occurrence of pancreatic hypertrophy and hyperplasia, not only in chicks but also in the young ages of different species.

Ileal digestibility of amino acids was evaluated on samples of broilers fed on diets containing raw, full-fat soybean and then the result showed that methionine had the highest digestibility value than others (Table 7). Therefore, this study confirmed that digestibility of sulphur-containing amino acids might not be affected by presence of TI in diets.

Furthermore, Batterham et al. (1993) reported that the growing pigs can tolerate the dietary levels of trypsin and chymotrypsin inhibitors, at least 4.7 and 4.5 mg/g, respectively; otherwise the weights of the liver and pancreas were significantly affected. In contrast to this, Struthers et al. (1983) reported that the TI neither from raw soy flour nor from any other soy products produced pancreatic enlargement in pigs or monkeys.

Although Grant (1989) reported that dietary soybean or TI do however altering the pancreatic secretion in humans, the effect of TI from soybean does not appear to be a potential hazard on human trypsin. Flavin (1982) also suggested that trypsin from human is more resistant to inhibition than is the trypsin of other mammalian species. Similarly, Baba et al. (1984) reported that major resection of the pancreas leads to disordering of the endocrine and exocrine pancreas. An investigation was conducted in identifying whether DNA concentration was proportionally increased due to the TI from soybean in samples taken from the pancreatic hypertrophy. Then, the results showed that the concentration of DNA or the ratio of tissue proteins to DNA was not significantly affected (Erdaw et al., 2017b).

Table 7. Effects of microbial protease and phytase, in diets, containing 75 g of raw soybean, on apparent ileal digestibility of crude protein and amino acids of broilers at d 24

Main effects		CP	Indispensable amino acids									Dispensable amino acids			
Protease Prot/kg	Phytase Fyt/kg		His	Arg	Thr	Val	Lys	Met	Ile	Leu	Phe	Ser	Gly	Pro	Ala
0		0.724^{b}	0.761^{b}	0.813^{b}	0.673^{b}	0.697^{b}	0.792^{c}	0.921	0.700^{b}	0.740^{b}	0.745	0.672	0.657^{b}	0.720^{b}	0.740^{b}
15000		0.718^{b}	0.769^{b}	0.815^{b}	0.676^{b}	0.719^{b}	0.802^{b}	0.923	0.696^{b}	0.737^{b}	0.741	0.665	0.674^{b}	0.723^{b}	0.741^{b}
~30000		0.745^{a}	0.788^{a}	0.839^{a}	0.705^{a}	0.748^{a}	0.834^{a}	0.917	0.758^{a}	0.791^{a}	0.790	0.722	0.683^{a}	0.743^{a}	0.776^{a}
	1000	0.734	0.768	0.821	0.678^{b}	0.717	0.801^{b}	0.923	0.718	0.757	0.760	0.684	0.672	0.731	0.756
	2000	0.730	0.772	0.822	0.678^{b}	0.719	0.810^{b}	0.922	0.725	0.764	0.767	0.696	0.678	0.735	0.757
	3000	0.723	0.778	0.825	0.698^{a}	0.728	0.816^{a}	0.916	0.716	0.751	0.753	0.684	0.664	0.722	0.747
Pooled SEM		0.004	0.006	0.003	0.005	0.005	0.004	0.002	0.006	0.005	0.005	0.006	0.005	0.004	0.004
Sources of variation															
Protease		***	***	***	**	***	***	NS	**	***	***	***	**	**	***
Phytase		NS	NS	*	NS	**	NS	NS	NS	0.09	NS	NS	NS	NS	NS
Protease x phytase		NS	NS	NS	NS	NS	NS	0.08	NS	**	*	NS	NS	NS	NS
Control‡		0.807	0.854	0.734	0.774	0.854	0.908	0.784	0.803	0.807	0.756	0.751	0.779	0.801	0.801
Contrasts															
Control‡ *vs* test-diet†		**	***	***	***	***	***	NS	***	***	***	***	***	***	***
Control‡ *vs* test-diet†+ pro-free		**	***	***	***	***	***	NS	***	***	***	***	***	***	***
Control‡ *vs* test-diet† +pro		*	**	**	**	**	**	NS	**	*	*	**	***	**	**
Test-diet† + pro-free *vs* test-diet† + pro		NS	NS	NS	NS	NS	NS	NS	NS	NS	NS	NS	NS	NS	NS

Source: Adopted from Erdaw et al., 2018. Control‡ = a diet without raw soybean; test-diet† = a diet containing 25 g of raw soybean; pro = protease.

13. Effects of Exogenous Protease Enzymes on Natural Protease Inhibitors

As shown in Table 8, due to supplementation of microbial feed enzymes (protease and phytase), weight of pancreas was varyingly decreased. Additionally, Adeyemo and Onilude (2013) found that supplementation of alpha-galactosidase enzyme reduced the ANF in soybeans and also enhanced its nutritional values. Microbial protease supplementation on pig diets improved the performance and this might be due to the implication that imposed on ANF (O'Doherty and Forde, 1999). Supplementation of protease enzymes helped to improve the performance and CP digestibility (Angel et al., 2011). Almost all enzymes currently being used to reduce the adverse effects of anti-nutritional factors that are varyingly found in many feed ingredients of mono-gastric animals (Bedford and Schulze, 1998).

Table 8. Effects of protease and phytase on pancreas of broilers (g/100 g of the body weight)

Main effects		Weight of the pancreas		
Protease, g/kg diet	**Phytase, g/kg diet**	**10 d**	**24 d**	**35 d**
0.0		0.72	0.37[a]	0.24
0.2		0.67	0.36[a]	0.24
0.4		0.71	0.25[b]	0.23
	0.1	0.73	0.33[a]	0.23
	0.2	0.69	0.34[a]	0.24
	0.3	0.67	0.31[b]	0.24
Pooled SEM	0.01	0.01	0.01	
Sources of variation				
Protease		NS	***	NS
Phytase		0.06	*	NS
Protease×phytase		NS	NS	NS

Sources: Adopted from Erdaw et al., 2018; d = day (s). A 25% of commercial soybean meal (equivalent to 75 g) was replaced by raw soybean.

Soybean meal (ESBM) had a lower concentration of antinutritional factors when it was either treated by fermentation or supplemented with feed enzymes (Stein, 2010).

A large variety of carbohydrases, proteases, phytases and lipases are used for these purposes are added to animal feed in order to increase digestibility to remove ANF and to improve nutrient availability, as well as for environmental reasons (Peek, 2009; Li et al., 2010).

Therefore, it was concluded that the increase in weight of the pancreas is due to hypertrophy than hyperplasia (Erdaw, 2016). Meaning pancreatic hypertrophy might be happened due to the cell-size increments than increase of the cell numbers.

CONCLUSION

One of the ANF that commonly and varyingly distributed throughout the nature is called protease inhibitors (trypsin inhibitor and chymotrypsin inhibitor). From which trypsin inhibitor is one of the adversely affecting protease inhibitors that are always presents /available in some important feeds or food ingredients of animals. Although TI is supposed to be responsible for pancreatic hypertrophy as well as inhibitions, uncertainty with regards to the biological threshold levels needs to be set out by conducting relevant experimentations. Since TI is highly concentrated in legume grains, and mainly located in the protein bodies of the cotyledon, strategic solution may be considered while planning to reduce the TI in feeds and diets. The contribution of BBI and to some extent KTI to the quality, particularly improving that of the contents of methionine and cysteine, needs to be considered in soybean based diets. Pancreases of all animals are not equally susceptible to being affected by the TI in their diets. It is recommended for further investigation about how to optimize reduction of the ANF and on the other hand avoiding that of denaturing those important nutrients.

REFERENCES

Adeola, O. & Cowieson, A. (2011). Board-invited review: Opportunities and challenges in using exogenous enzymes to improve no-nruminant animal production. *Journal of Animal Sciences*, *89*, 3189-3218.

Adeyemo, S. M. & Onilude, A. A. (2013). Enzymatic reduction of anti-nutritional factors in fermenting soybeans by *Lactobacillus plantarum* isolates from fermenting cereals. *Nigerian Food Journal*, *31*, 84-90.

Anderson, R. L. & Wolf, W. J. (1995). Compositional changes in trypsin inhibitors, phytic acid, saponins and isoflavones related to soybean processing. *The Journal of Nutrition*, *125*, 581S-588S.

Akande, K. E., Doma, U. D., Agu, H. O. & Adamu, H. M. (2010). Major antinutrients found in plant protein sources: their effect on nutrition. *Pakistan Journal of Nutrition*, *9*, 827-832.

Anderson, R. L. & Wolf, W. J. (1995). Compositional changes in trypsin inhibitors, phytic acid, saponins and isoflavones related to soybean processing. *The Journal of Nutrition*, *125*(suppl_3), 581S-588S.

Angel, C. R., Saylor, W., Vieira, S. L. & Ward, N. (2011). Effects of a monocomponent protease on performance and protein utilization in 7 to 22 -day-old broiler chickens. *Poultry Science*, *90*, 2281-2286.

Anderson, R. L. (1986). Trypsin inhibitors of raw and heat-treated soybean meals (Anhydrotrypsin, Affinity Chromatography). http://digital commons.unl.edu/dissertations/AAI8629524.

Ao, T. (2011). Using exogenous enzymes to increase the nutritional value of soybean meal in poultry diet. In: Soybean and nutrition, edited by El-Shemy, H. InTech, Rijeka, Croatia, pp. 201-214.

Baba, N., Suzuki, T. & Tobe, T. (1984). Effects of trypsin inhibitor on the remnant pancreas after 85 percent pancreatectomy in rats. *The Japanese Journal of Surgery*, *14*, 420-431.

Barletta, A. (2011). Introduction: current market and expected developments. In: Enzymes in Farm Animal Nutrition, edited by Bedford. M. R., and Partridge, G. G. 1-11. CABI, Wallingford, UK.

Baker, E. C. & Rackis, J. J. (1986). Preparation of unheated soy protein isolates with low trypsin inhibitor content. In: *Nutritional and*

Toxicological Significance of Enzyme Inhibitors in Foods, (349-355), edited by Friedman M. Springer, Boston, MA, USA.

Batterham, E. S., Saini, H. S., Andersen, L. M. & Baigent, R. D. (1993). Tolerance of growing pigs to trypsin and chymotrypsin inhibitors in chickpeas (*Cicer arietinum*) and pigeonpeas (*Cajanus cajan*). *Journal of the Science of Food and Agriculture*, *61*, 211-216.

Bedford, M. A. & Schulze, H. (1998). Exogenous enzymes for pigs and poultry. *Nutrition research reviews*, *11*, 91-114.

Bin, P., Huang, R. & Zhou, X. (2017). Oxidation resistance of the sulfur amino acids: methionine and cysteine. *BioMed Research International*, 2017, Article ID 9584932, 6 pages. Available online from: https://www.hindawi.com/journals/bmri/2017/9584932/.

Brandon, D. L., Bates, A. H. & Friedman, M. (1991). ELISA analysis of soybean trypsin inhibitors in processed foods. In: *Nutritional and toxicological consequences of food processing*, (pp. 321-337), edited by Friedman M. Springer US.

Broadway, R. M. & Duffey, S. S. (1986). Plant proteinase inhibitors: mechanism of action and effect on the growth and digestive physiology of larval *Heliothis zea* and *Spodoptera exiqua*. *Journal of Insect Physiology*, *32*, 827-833.

Brosnan, J. T. & Brosnan, M. E. (2006). The sulphur-containing amino acids: an overview. *The Journal of Nutrition*, *136*, 1636S-1640S.

Burns, R. A. (1987). Protease inhibitors in processed plant foods. *Journal of Food Protection*, *50*, 161-166.

Carvalho, M. R. B. & Sgarbieri, V. C. (1997). Heat treatment and inactivation of trypsin-chymotrypsin inhibitors and lectins from beans (*Phaseolus vulgaris* L.). *Journal of Food Biochemistry*, *21*, 219-233.

Carvalho, A. W. D., Natal, D. I. G., Silva, C. O. D., Dantas, M. I. D. S., Barros, E. G. D., Ribeiro, S. M. R. & Martino, H. S. D. (2013). Heat-treatment reduces anti-nutritional phytochemicals and maintains protein quality in genetically improved hulled soybean flour. *Food Science and Technology*, *33*, 310-315.

Chen, Y., Duan, W., Wang, L., Zhang, S. & Zhou, Y. (2013). Efects of thermostable phytase supplementation on the growth performance and nutrient digestibility of broilers. *International Journal of Poultry Science*, *12*, 441-444.

Clarke, E. & Wiseman, J. (2007). Effects of extrusion conditions on trypsin inhibitor activity of full fat soybeans and subsequent effects on their nutritional value for young broilers. *British Poultry Science*, *48*, 703-712.

Clarke, E. & Wiseman, J. (2005). Effects of variability in trypsin inhibitor content of soya bean meals on true and apparent ileal digestibility of amino acids and pancreas size in broiler chicks. *Animal Feed Science and Technology*, *121*, 125-138.

Clarke, E. & Wiseman, J. (2000). Developments in plant breeding for improved nutritional quality of soya beans II. Anti-nutritional factors. *The Journal of Agricultural Science*, *134*, 125-136.

Clemente, A., Arques, M. C., Dalmais, M., Le Signor, C., Chinoy, C., Olias, R. & Domoney, C. (2015). Eliminating anti-nutritional plant food proteins: the case of seed protease inhibitors in pea. *PloS One*, *10*, e0134634.

Cowieson, A. J. & Ravindran, V. (2007). Effect of phytic acid and microbial phytase on the flow and amino acid composition of endogenous protein at the terminal ileum of growing broiler chickens. *British Journal of Nutrition*, *98*, 745-752.

Dale, N. (1996). Variation in feed ingredient quality: oilseed meals. *Animal Feed Science Technology*, *59*, 129-135.

Daneluti, A. L. M. & Matos, J. D. R. (2013). Study of thermal behavior of phytic acid. *Brazilian Journal of Pharmaceutical Sciences*, *49*, 275-283.

de-Carvalho, W. L., de-Almeida-Oliveira, M. G., de Barros, E. G. & Moreira, M. A. (1998). Determination of genotypic classes for trypsin inhibitor in soybean seeds. *Biotechnology Techniques*, *12*, 859-863.

Dourado, L. R. B., Pascoal, L. A. F., Sakomura, N. K., Costa, F. G. P. & Biagiot, T. I. D. (2011). Soybeans (Glycine max) and soybean products in poultry and swine nutrition, In: Recent Trends for Enhancing the Diversity and Quality of Soybean Products, edited by Dora Krezhova. InTech, Rijeka, Croatia, pp. 175-190.

Erdaw, M. M., Perez-Maldonado, R. A. & Iji, P. A. (2018). Supplementation of broiler diets with high levels of microbial protease and phytase enables partial replacement of commercial soybean meal with raw, full-fat soybean. *Journal of Animal Physiology and Animal Nutrition*, *102*, 755-768.

Erdaw, M. M., Perez-Maldonado, R. A., Bhuiyan, M. & Iji, P. A. (2017a). Partial replacement of commercial soybean meal with raw, full-fat soybean meal supplemented with varying levels of protease in diets of broiler chickens. *South African Journal of Animal Science*, *47*, 61-71.

Erdaw, M. M., Wu, S. & Iji, P. A. (2017b). Growth and physiological responses of broiler chickens to diets containing raw, full-fat soybean meal and supplemented with a high-impact protease. *Asian-Australas Journal of Animal Sciences*, *30*, 1303-1313.

Erdaw, M. M. (2016). Maximizing the nutritional value of unprocessed soybean meal through supplementation with complex microbial enzyme products. PhD thesis submitted to the University of New England, Australia.

El-Sayed Embaby, H. (2010). Effect of heat treatments on certain antinutrients and *in vitro* protein digestibility of peanut and sesame seeds. *Food Science and Technology Research*, *17*, 31-38.

Ezeagu, I. E. (2006). Efficiency of inactivation of trypsin inhibitory activity in some selected tropical plant seeds by autoclaving. *Biochememistry*, *1*, 21-24.

Flavin, D. F. (1982). The effects of soybean trypsin inhibitors on the pancreas of animals and man: a review. *Veterinary and Human Toxicology*, *24*, 25-28.

Francis, G., Makkar, H. P. & Becker, K. (2001). Antinutritional factors present in plant-derived alternate fish feed ingredients and their effects in fish. *Aquaculture*, *199*, 197-227.

Friedman, M. & Brandon, D. L. (2001). Nutritional and health benefits of soy proteins. *Journal of Agricultural and Food Chemistry*, *49*, 1069-1086.

Frost, A. B. & Mann, G. V. (1966). Effects of cystine deficiency and trypsin inhibitor on the metabolism of methionine. *The Journal of Nutrition*, *89*, 49-54.

García-Rebollar, P., Cámara, L., Lázaro, R. P., Dapoza, C., Pérez-Maldonado, R. & Mateos, G. G. (2016). Influence of the origin of the beans on the chemical composition and nutritive value of commercial soybean meals. *Animal Feed Science and Technology*, *221*, 245-261.

Gilani, G. S., Cockell, K. A. & Sepehr, E. (2005). Effects of antinutritional factors on protein digestibility and amino acid availability in foods. *Journal of AOAC International*, *88*, 967-987.

Gilani, G. S., Xiao, C. W. & Cockell, K. A. (2012). Impact of antinutritional factors in food proteins on the digestibility of protein and the bioavailability of amino acids and on protein quality. *British Journal of Nutrition*, *108*, S315-S332.

Grant, G. (1989). Anti-nutritional effects of soybean: a review. *Progress in Food and Nutrition Science*, *13*, 317-348.

Ge, Y. C. & Morgan, R. G. H. (1993). The effect of trypsin inhibitor on the pancreas and small intestine of mice. *British Journal of Nutrition*, *70*, 333-345.

Griffiths, D. W. (1984). The trypsin and chymotrypsin inhibitor activities of various pea (*Pisum* spp.) and field bean (*Vicia faba*) cultivars. *Journal of the Science of Food and Agriculture*, *35*, 481-486.

Guggenbuhl, P., Waché, Y. & Wilson, J. W. (2012). Effects of dietary supplementation with a protease on the apparent ileal digestibility of the weaned piglet. *Journal of Animal Science*, *90*, 152-154.

Guillamon, E., Pedrosa, M. M., Burbano, C., Cuadrado, C., de Cortes Sánchez, M. & Muzquiz, M. (2008). The trypsin inhibitors present in seed of different grain legume species and cultivar. *Food Chemistry*, *107*, 68-74.

Hajos, G., Gelencser, E., Pusztai, A., Grant, G., Sakhri, M. & Bardocz, S. (1995). Biological effects and survival of trypsin inhibitors and the agglutinin from soybean in the small intestine of the rat. *Journal of Agricultural and Food Chemistry*, *43*, 165-170.

House, W. A., Van Campen, D. R. & Welch, R. M. (1996). Influence of dietary sulfur-containing amino acids on the bioavailability to rats of zinc in corn kernels. *Nutrition Research*, *16*, 225-235.

Jordão, A. J. & Marchin, J. S. (1998). Methionine supplementation of soya products: effects on nitrogen balance parameters. *Archivos Latinoamericanos de Nutrición*, *48*, 35-40.

Kanjanapruthipong, J. (1998). Supplementation of milk replacers containing soy protein with threonine, methionine, and lysine in the diets of calves. *Journal of Dairy Science*, *81*, 2912-2915.

Karr-Lilienthal, L. K., Grieshop, C. M., Merchen, N. R., Mahan, D. C. & Fahey, G. C. (2004). Chemical composition and protein quality comparisons of soybeans and soybean meals from five leading soybean-producing countries. *Journal of Agricultural and Food Chemistry*, *52*, 6193-6199.

Kaur, S., Sharma, S., Dar, B. N. & Singh, B. (2012). Optimization of process for reduction of antinutritional factors in edible cereal brans. *Food Science and Technology International*, *18*, 445-454.

Khokhar, S. & Apenten, R. K. O. (2003). Antinutritional factors in food legumes and effects of processing. In The role of food, agriculture, forestry and fisheries in human nutrition. Encyclopedia of Life Support Systems (EOLSS), Co Ltd., Oxford, UK, vol *1*, chapter 4, pp. 82-116.

Kim, G., Weiss, S. J. & Levine, R. L. (2014). Methionine oxidation and reduction in proteins. *Biochimica et Biophysica Acta (BBA)-General Subjects*, *1840*, 901-905.

Kochhar, N., Walker, A. F. & Pike, D. J. (1988). Effect of variety on protein content, amino acid composition and trypsin inhibitor activity of cowpeas. *Food Chemistry*, *29*, 65-78.

Krishnan, H. B., Bennett, J. O., Kim, W. S., Krishnan, A. H. & Mawhinney, T. P. (2005). Nitrogen lowers the sulfur amino acid content of soybean (*Glycine max* [L.] Merr.) by regulating the accumulation of Bowman-Birk protease inhibitor. *Journal of Agricultural and Food Chemistry*, *53*, 6347-6354.

Krogdahl, Å., Lea, T. B. & Olli, J. J. (1994). Soybean proteinase inhibitors affect intestinal trypsin activities and amino acid digestibilities in rainbow trout (*Oncorhynchus mykiss*). *Comparative Biochemistry and Physiology Part A: Physiology*, *107*, 215-219.

Krogdahl, A. & Berg-Lea, T. (1992). Effects of a soybean proteinase inhibitor on trypsin activity and digestibilities of amino acids in rainbow trout measured in the proximal and distal intestine and in faeces. *Aquaculture*, *100*, 232-233.

Kwong, E. & Barnes, R. H. (1963). Effect of soybean trypsin inhibitor on methionine and cystine utilization. *Journal of Nutrition*, *81*, 392-398.

Laskowski, M. & Kato, I. (1980). Protein inhibitors of proteinases. *Annual Review of Biochemistry*, *49*, 593-626.

Levine, R. L., Mosoni, L., Berlett, B. S. & Stadtman, E. R. (1996). Methionine residues as endogenous antioxidants in proteins. *Proceedings of the National Academy of Sciences*, *93*, 15036-15040.

Liener, I. E. & Hasdai, A. (1986). The effect of the long-term feeding of raw soyflour on the pancreas of the mouse and hamster. In: Nutritional and Toxicological Significance of Enzyme Inhibitors in Foods, edited by Friedman M. pp. 189-197. Springer, Boston, MA.

Lesake, K. L., Zhang, B. & Coon, C. N. (1995). The use of low alpha-glucoside protein products as a protein source in chicken diets. *Animal Feed Science and Technology*, *54*, 275-286.

Li, Q., Yi, L., Marek, P. & Iverson, B. L. (2013). Commercial proteases: present and future. *FEBS Letters*, *587*, 1155-1163.

Li, Y., Fang, Z., Dai, J., Partridge, G., Ru, Y. & Peng, J. (2010). Corn extrusion and enzyme addition improves digestibility of corn/soy based diets by pigs: in vitro and in vivo studies. *Animal feed science and technology*, *158*, 146-154.

Li, D. F., Nelssen, J. L., Reddy, P. G., Blecha, F., Hancock, J. D., Alee, G. L., Goodband, R. D. & Klemm, R. D. (1990). Transient hypersensitivity to soybean meal in the early-weaned pig. *Journal of Animal Science*, *68*, 1790-1759.

Livingstone, D., Beilinson, V., Kalyaeva, M., Schmidt, M. A., Herman, E. M. & Nielsen, N. C. (2007). Reduction of protease inhibitor activity by expression of a mutant Bowman-Birk gene in soybean seed. *Plant Molecular Biology*, *64*, 397-408.

Liener, I. E. (1994). Implications of antinutritional components in soybean foods. *Critical Reviews in Food Science and Nutrition*, *34*, 31-67.

Luo, Y. W. & Xie, W. H. (2013). Effect of different processing methods on certain antinutritional factors and protein digestibility in green and white faba bean (*Vicia faba* L.). *CyTA-Journal of Food*, *11*, 43-49.

McDonald, P., Greenhalgh, J. F. D., Edwards, R. & Morgan, C. A. (2002). Animal Nutrition. 6th edition. Pearson Education Limited, Edinbutgh Gate Harlow Essex CM20 2JE, England., pp. 543.

Ma, W., Tang, C. & Lai, L. (2005). Specificity of trypsin and chymotrypsin: loop-motion-controlled dynamic correlation as a determinant. *Biophysical Journal*, *89*, 1183-1193.

Makkar, H. P. S., Siddhuraju, P. & Becker, K. (2007). Plant secondary metabolites: In Trypsin Inhibitor, edited by John M. Walker. *E-Book.*, *393*, pp. 1-6. Humana Press.

Marie, J. (2017). *Food Sources of Methionine and Cysteine.* Livestrong.com. Available online from: https://www.livestrong.com/article/274583-food-sources-of-methionine-cysteine/.

Marsman, G. J., Gruppen, H., Van der Poel, A. F., Kwakkel, R. P., Verstegen, M. W. & Voragen, A. G. (1997). The effect of thermal processing and enzyme treatments of soybean meal on growth performance, ileal nutrient digestibilities, and chyme characteristics in broiler chicks. *Poultry Science*, *76*, 864-872.

Mikić, A., Perić, V., Đorđević, V., Srebrić, M. & Mihailović, V. (2009). Anti-nutritional factors in some grain legumes. *Biotechnology in Animal Husbandry*, *25*, 1181-1188.

Mulimani, V. H. & Paramjyothi, S. (1994). Effect of heat treatments on trypsin/chyomotrypsin inhibitor activity of red gram (*Cajanus cajan* L.). *Plant Foods for Human Nutrition*, *46*, 103-107.

Murray, D. D. (2009). Use of canola in aquafeeds. University of Saskatchewan, Department of Animal and Poultry Science. Canada.

Nabil, S. (2014). *The Role of Anti-Nutritional Factors in Seeds & Legumes.* Accessible from: https://www.insta-pro.com/en/2014/04/01/ the-role-of-anti-nutritional-factors-in-seeds-legumes/.

Neus, J. D., Fehr, W. R. & Schnebly, S. R. (2005). Agronomic and seed characteristics of soybean with reduced raffinose and stachyose. *Crop Science*, *45*, 589-592.

Newkirk, R. (2010). Soybean. Feed industry guide, 1st edition. Canadian Int'l Grains Institute, pp. 48.

O'Doherty, J. V. & Forde, S. (1999). The effect of protease and α-galactosidase supplementation on the nutritive value of peas for growing and finishing pigs. *Irish Journal of Agricultural and Food Research*, *38*, 217-226.

Oke, D. B. (2007). Mechanism of action, toxicity and nutritional significance of heat-labile antinutritional factors in some legumes: a review. *Journal of Food and Technology*, *5*, 286-289.

Osborne, T. B. & Mendel, L. B. (1917). The use of soybeans as food. *Journal of Biological Chemistry*, *32*, 369-387.

Paiva, K. C., Kelli, C. P., Barbieri de-Carvalho, M. R. & Pizauro, J. M. (2011). Effectiveness inactivation of trypsin inhibitor from brazilian cultivars of beans (*Phaseolus vulgaris* L.). *Alimentos e Nutriçao*, *22*, 331-337.

Peek, H. W., Van der Klis, J. D., Vermeulen, B. & Landman, W. J. M. (2009). Dietary protease can alleviate negative effects of a coccidiosis infection on production performance in broiler chickens. *Animal Feed Science and Technology*, *150*, 151-159.

Perić, V., Žilić, S., Srebrić, M. & Mikić, A. (2011). Nutritional value of the grain of Kunitz-free soybean cultivars. *Biotechnology in Animal Husbandry*, *27*, 1537-1542.

Pettersson, D. & Pontoppidan, K. (2013). Soybean meal and the potential for upgrading its feeding value by enzyme supplementation. In: *Soybean - Bio-Active Compounds*, edited by Hany A. El-Shemy., pp. 288-307. Intech, Open Access Publisher.

Rackis, J. J. (1981). Significance of soya trypsin inhibitors in nutrition. *Journal of the American Oil Chemists' Society*, *58*, 495-501.

Rackis, J. J., McGhee, J. E. & Booth, A. N. (1975). Biological threshold levels of soybean trypsin inhibitors by rat bioassay. *Cereal Chemistry*, *52*, 85-92.

Rackis, J. J., McGee, J. E., Gumbmann, M. R. & Booth, A. N. (1979). Effects of soy proteins containing trypsin inhibitors in long term feeding studies in rats. *Journal of the American Oil Chemists' Society*, *56*, 162-168.

Rackis, J. J., Wolf, W. J. & Baker, E. C. (1986). Protease inhibitors in plant foods: content and inactivation. In: *Nutritional and Toxicological Significance of Enzyme Inhibitors in Foods*, (pp. 299-347), edited by Friedman M. Springer USA.

Ramadoss, B. R. & Shunmugam, A. S. (2014). Anti-dietetic factors in legumes-local methods to reduce them. *International Journal of Food and Nutritional Science*, *3*, 84-89.

Ravindran, V., Abdollahi, M. & Bootwalla, S. (2014a). Nutrient analysis, apparent metabolisable energy and ileal amino acid digestibility of full fat soybean for broilers. *Animal Feed Science and Technology*, *197*, 233-240.

Ravindran, V., Abdollahi, M. R. & Bootwalla, S. M. (2014b). Nutrient analysis, metabolizable energy, and digestible amino acids of soybean meals of different origins for broilers. *Poultry Science*, *93*, 2567-2577.

Roos, N., Sørensen, J. C., Sørensen, H., Rasmussen, S. K., Briend, A., Yang, Z. & Huffman, S. L. (2013). Screening for anti-nutritional compounds in complementary foods and food aid products for infants and young children. *Maternal and Child Nutrition*, *9*, 47-71.

Roux, M. (2016). *How animal feed enzymes can improve producer profitability.* Retrieved from: https://www.wattagnet.com/articles/25854-how-animal-feed-enzymes-can-improve-producer-profitability.

Ruiz, N. & de Belalcázar, F. (2005). Field observation: Trypsin inhibitors in soybean meal are correlated with outbreaks of feed passage in broilers. *Poultry Science*, *84*, 70-70.

Sasaki, J. & Nakashima, N. (2000). Methionine-independent initiation of translation in the capsid protein of an insect RNA virus. *Proceedings of the National Academy of Sciences*, *97*, 1512-1515.

Sessa, D. J. & Bietz, J. A. (1986). Toasted soybean flour components with trypsin inhibitor activity. *Journal of the American Oil Chemists' Society*, *63*, 784-788.

Sousa, D. O., Carvalho, A. F., Oliveira, J. T. A., Farias, D. F., Castelar, I., Oliveira, H. P. & Vasconcelos, I. M. (2015). Increased levels of antinutritional and/or defense proteins reduced the protein quality of a disease-resistant soybean cultivar. *Nutrients*, *7*, 6038-6054.

Soetan, K. & Oyewol, O. (2009). The need for adequate processing to reduce the antinutritional factors in plants used as human foods and animal feeds: A review. *African Journal of Food Science*, *3*, 223-232.

Stein, H. H. (2010). Nutrition advances benefit swine and the bottom line. Pages 10-13 in *Feed Management*. Accessed online from: https:// www.wattagnet.com/ articles/ 5670- nutrition- advances- benefit-swine-and-the-bottom-line.

Struthers, B. J., Macdonald, J. R., Dahlgren, R. R. & Hopkins, D. T. (1983). Effects on the monkey, pig and rat pancreas of soy products with varying levels of trypsin inhibitor and comparison with the administration of cholecystokinin. *The Journal of Nutrition*, *113*, 86-97.

Sumantha, A., Larroche, C. & Pandey, A. (2006). Microbiology and industrial biotechnology of food-grade proteases: A perspective. *Food Technology and Biotechnology*, *44*, 211-220.

Tacon, A. G. J. (1985). *Morphological signs of nutrient deficiency and toxicity in farmed fish.* Aquaculture Development and Coordination Program. UN. FAO.

Tan-Wilson, A. L. & Wilson, K. A. (1986). Relevance of multiple soybean trypsin inhibitor forms to nutritional quality. In: *Nutritional and toxicological significance of enzyme inhibitors in foods*, (pp. 391-411), edited by Friedman M. Springer US.

Teeter, R. G., Baker, D. H. & Corbin, J. E. (1978). Methionine and cystine requirements of the cat. *The Journal of Nutrition*, *108*, 291-295.

Van Eys, J. E., Offner, A. & Bach, A. (2004). Manual of quality analyses for soybean products in the feed industry. Page 105. American Soybean Association, St. Louis, MO.

Varga-Visi, É., Pohn, G., Albert, C., Mándoki, Z. & Csapó, J. (2009). The effect of thermic treatment conditions on the amino acid composition of soybean and maize. *Acta Universitatis Sapientiae, Alimentaria*, *2*, 166-173.

Vidal-Valverde, C., Frias, J., Diaz-Pollan, C., Fernandez, M., Lopez-Jurado, M. & Urbano, G. (1997). Influence of processing on trypsin inhibitor activity of faba beans and its physiological effect. *Journal of Agricultural and Food Chemistry*, *45*, 3559-3564.

Vogt, W. (1995). Oxidation of methionyl residues in proteins: tools, targets, and reversal. *Free Radical Biology and Medicine*, *18*, 93-105.

Woyengo, T. A. & Nyachoti, C. M. (2011). Review: Supplementation of phytase and carbohydrases to diets for poultry. *Canadian Journal of Animal Science*, *91*, 177-192.

Zhou, J., Liu, C. & Tsou, C. (1989). Kinetics of trypsin inhibition by its specific inhibitors. *Biochemistry*, *28*, 1070-1076.

In: Proteases: Functions, Mechanisms and Uses ISBN: 978-1-53615-854-0
Editor: Nelson Pérez Guerra

Chapter 8

BIOINFORMATICS RESOURCES FOR PROTEASE ANALYSIS

Anoushka Kotra and Neetu Jabalia*, PhD
Amity Institute of Biotechnology, Amity University Uttar Pradesh, Noida, India

ABSTRACT

Bioinformatics is an interdisciplinary area that is associated with the development of various tools for analysis and interpretation of biological data. One of the major objectives of this chapter is to provide a better understanding of protease and its functions at the molecular level by using bioinformatics approaches including databases, tools, visualization and data analysis. Proteases are responsible for proteolysis of proteins, one of the major group of industrial enzymes involved in digestion, conception, birth, growth, maturation, aging and even death of all organisms. In addition to food industry, use of proteases in pharmaceutics is gaining more attention as several clinical studies are indicating their applications in various diseases. Therefore, the major focus of the present chapter will include many applications of tools and databases that are available for providing information on proteases like MEROPS, Gypsy Database, Knottin database, HIV protease database and PLANT-PI. The chapter

* Corresponding Author's E-mail: njabalia@gmail.com.

gives a detailed overview of the significant database which provides information on enzymes such as BRENDA, KEGG and CSA. The present chapter also highlights ExploreEnz, IntEnz, EzCat database, SFLD and CPDB. The last section describes the tools like PROSPER, Peptide Cutter, PSORT, for enzyme analysis which will help the readers to effectively use the information for their research endeavors.

Keywords: bioinformatics, protease, MEROPS, BRENDA, enzymes

1. Introduction

Proteases (proteolytic enzymes), a major class of hydrolytic enzymes are ubiquitous in nature and of immense importance in vital biological processes. They can perform specific changes in proteins by restricted cleavage, e.g., activation of zymogenic types of enzymes, blood coagulating and fibrin lysis along with handling of secretory proteins transportation across the membranes [1]. Proteases are commercially exploited extensively in detergent, food, leather, pharmaceutical, diagnostics, fine chemical industries and other industrial sectors [2]. Additionally, they are used in biological processes such as metabolism regulation, modification of enzymes, programmed cell death pathways and complement system.

Proteases constitute a huge collection of enzymes that conduct protein degradation through hydrolysis of peptide bond between amino acids in a polypeptide chain [3]. Although, proteases from microbial sources (bacterial and fungal) playing vital role in various industrial applications are most commonly found because of their rapid growth and ease in availability [4], plants are also potential source of proteases, being involved in biological processes such as releasing peptide hormones, cleaving transcription factors, proenzymes activation and specific protein degradation. According to report by Global Trends and Forecasts, market worth of protease enzymes is expected to be 2,767 Million dollars by the year 2019. Proteases have gained attention worldwide due to the specificity of their action in various physiological and biological applications.

There are various biological sources of proteases. These include:

- *Plant proteases:* Plants counter to herbivore attack by protection against proteases that hinder the digestive proteases of the herbivores that feed on plants and also cause a drop in their growth, development and fertility [5]. The extraction of protease from plants is a time consuming process and is based upon a number of parameters such as the land area for appropriate growth and other physiological parameters. The most commonly derived plant proteases are: papain, keratinases and bromelain [1].
- *Animal proteases:* The production of animal proteases depends upon various political and agricultural policies that further govern the availability of livestock for slaughter. The most commonly derived animal proteases are trypsin, pepsin, chymotrypsin and rennins.
- *Proteases from microbes:* Proteases of microbial origin are abundantly used due to their ability to alter the genetic makeup and their involvement in a variety of biochemical reactions. They are better than animal or plant proteases since they have all the characteristics required for biotechnological applications. Different microbes yield different proteases.
- *Bacteria:* Most of the commercial proteases are obtained from bacteria. They may be alkaline or neutral. Neutrase is a neutral protease that is widely used in the brewing industry [2].
- *Fungi:* The fungi proteases are highly specific to the substrate and are active over a wide pH range (4-11). *Aspergillus oryzae* yield acid, neutral and alkaline proteases. Most of the fungi proteases find their application in the cheese industry [3].
- *Viruses:* The viral proteases are useful for curing certain fatal diseases such as AIDS, cancer, etc., due to their functional involvement [4].

1.1.Types of Proteases

Proteases can be grouped into two broad categories. These categories depend upon the nature of first amino acid and the cleaving site of peptide bond and their site of action. The 2 major groups are: exopeptidases and endopeptidases (Figure 1).

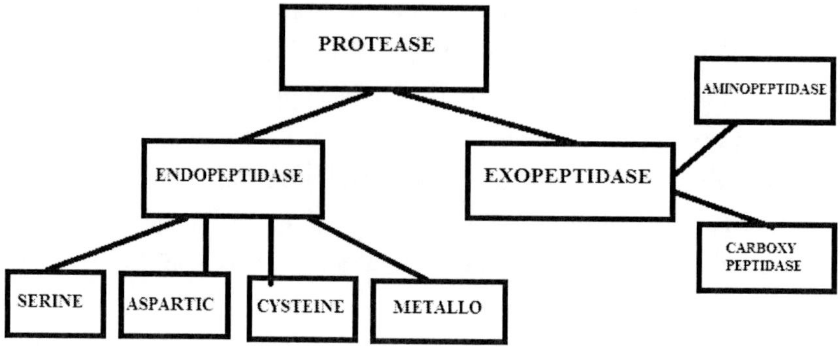

Figure 1. Classification of protease based on peptide cleavage.

1.1.1. Exopeptidases

They cleave at the end termini of a polypeptide chain i.e., N terminus or C terminus. They are of two types: Aminopeptidase and Carboxy-peptidase.

1.1.1.1. Aminopeptidase

These enzymes cleave at the N end of the polypeptide chain and may release one, two or three amino acids. The enzymes are abundantly present in microbial species which include bacteria and fungi.

1.1.1.2. Carboxypeptidase

They cleave at the C end of the protein and may release one or two amino acids. Carboxypeptidase is further categorized into 3 types: Serine carboxypeptidase, Metallocarboxypeptidase and Cysteine carboxy-peptidase. They vary in their different types of properties such as pH, molecular weight, presence of Zn^{2+}, etc.

1.1.2. Endopeptidase

Unlike exopeptidase that cleave at the terminus of the polypeptide chain, these enzymes cleave within the polypeptide chain. They are further categorized into 4 subgroups: Serine proteases, Subtilisins, Aspartic proteases, Cysteine/Thiol proteases and Metalloproteases.

1.1.2.1. Serine Proteases

Serine molecule is present in the active domain of the enzyme. Serine proteases are present in viruses, bacteria and other eukaryotes. Optimum pH activity of these enzymes is around 7-11.

1.1.2.2. Aspartic Proteases

They are also known as acidic proteases and are based upon the catalytic activity of aspartic residues. They function at acidic pH ranging from 3.0 to 4.5.

1.1.2.3. Cysteine/Thiol Proteases

Their activity is based upon the catalytic dyad of cysteine and histidine residues. They are found both in prokaryotes and eukaryotes. They cause hydrolysis of carboxylic acid derivatives [5].

1.1.2.4. Metalloproteases

Their activity is governed by the presence of a divalent metal ion. They are present in snake venoms, bacteria and other higher organisms in different forms. They are further classified as neutral, alkaline, *Myxobacter* I and *Myxobacter* II.

1.2. Applications of Proteases

1.2.1. Food Industry

Enzymes can be applied for various aspects in the food industry such as cheese making, baking, meat tenderization, etc. Papain is used in fruit juices.

1.2.2. Detergent Industry

The proteases used in detergent industry must have broad specificity in order to carry out the removal of different kinds of stains. The ionic strength of the protease is an important parameter to be considered for its use in the detergent industry. The most commonly used proteases in the detergent industry are derived from Bacillus strains which are serine proteases.

1.2.3. Dairy Industry

Proteases may be used in cheese making. Enzymes like rennin and chymosin are the ones that are most commonly used. Chymosin is preferred due to its highly specific nature for casein. Various other milk-coagulating enzymes can also be used.

1.2.4. Leather Industry

There has been a significant improvement in the quality of the leather by replacing chemicals with the use of enzymes that have also contributed to environmental pollution control. The proteases are used for selective hydrolysis of skin in pre-tanning operations. Alkaline proteases combined with lime and sodium chloride may also be used for dehairing purpose in order to reduce the generation of wastewater.

1.2.5. Brewing Industry

Proteases used in brewing industry act as a source of nitrogen required by the microbes during fermentation process. These are usually worth *Bacillus subtilis* derived proteases that solubilize proteins in barley to release free peptides and amino acids.

1.2.6. Pharmaceutical Industry

Many proteases such as Nortis from *Aspergillus oryzae* have been administered orally to treat lytic enzyme deficiency syndrome.

1.2.7. Therapeutics

Various proteases have been used in the form of injections to cause reduction in tissue inflammation and pain. Some of these enzymes also aid in digestion [6,7,8].

1.3. Role of Bioinformatics Studies in Protease

Bioinformatics is an interdisciplinary branch of science that combines software applications and algorithms to analyze biological information. Since it is difficult to analyze biological data at the molecular level, this branch aims to simplify its analysis by combining informatics, applied mathematics, computer science, chemistry, biochemistry, etc. (Figure 2). The primary focus lies on alignment of sequences (MSA- Multiple Sequence Alignment, etc.), gene discovery, protein sequencing, studying protein-protein interactions, gene assembly.

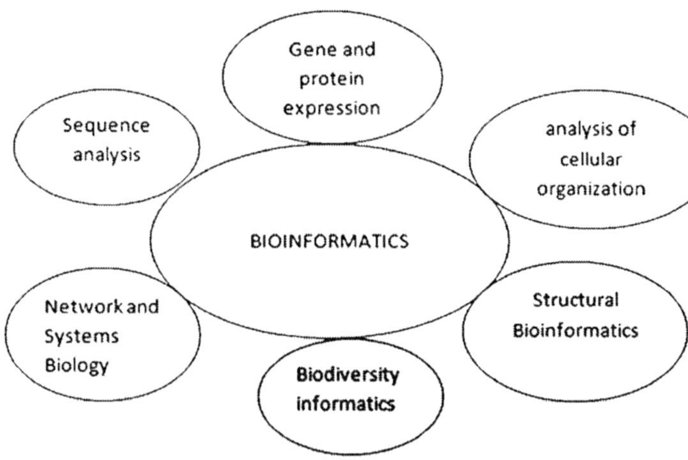

Figure 2. Various applications of Bioinformatics.

Bioinformatics is important to understand various biological processes in a diversity of organisms. Over a few years, a lot of information has been generated regarding molecular biology due to increasing development in genomics, proteomics and other molecular related technologies. Thus, these mathematical and computer approaches used for the understanding of biological processes is now known as "Bioinformatics".

Different areas within Computational Biology and Bioinformatics include:

1. Developing and utilizing computer programs that help in using, managing and analyzing various types of information.
2. Developing new programs, mathematical formulas and applying statistics to determine the relationship between large groups of data. For instance, various techniques are available to determine the position of a gene within a particular sequence, databases to identify 3D structure of a protein and clustering protein sequences into families of related sequences.

The Human Genome Project and advancement in DNA Sequencing Technology lead to a tremendous growth in the field of bioinformatics [9].

Biomolecular science has completely revolutionized with the onset of bioinformatics, sequencing and structural biology. Millions of sequences are available in Uniprot, and thousands of protein 3D structures are available in PDB (Protein Data Bank). A large number of softwares are freely available online to analyze an even larger quantity of biological data. The computation may take place either locally or online i.e., over the internet due to increase in hardware and network capabilities.

The most important aspect of an enzyme is its chemical mechanism by which it functions. Thus, data related to enzymes is of prime importance to the scientists. Bioinformatics provide different tools and databases that are mostly free for the analysis of enzymes. Some of these are BRENDA to describe the chemistry, specificity to a substrate, kinetics and various biological sources of enzymes, KEGG for determining the metabolism of enzymes, EzCatDB, SFLD and MACiE are important archives to provide information on chemical mechanisms by which the enzymes function (Figure 3).

Various *in silico* approaches have been derived from Chem-informatics to study the interaction of drugs, toxins and other proteins with biological system. The energy of enzyme catalyzed chemical reaction can be understood with the help of quantum and chemical methods such as DFT and QM/MM. Thus, computational techniques that combine machine learning, with the help of bioinformatics and cheminformatics, to quantum chemistry is required for interpretation and analysis of the sheer quantity of enzymatic data for its commercial application in areas of food technology, pharmaceuticals, medical and therapeutics [10].

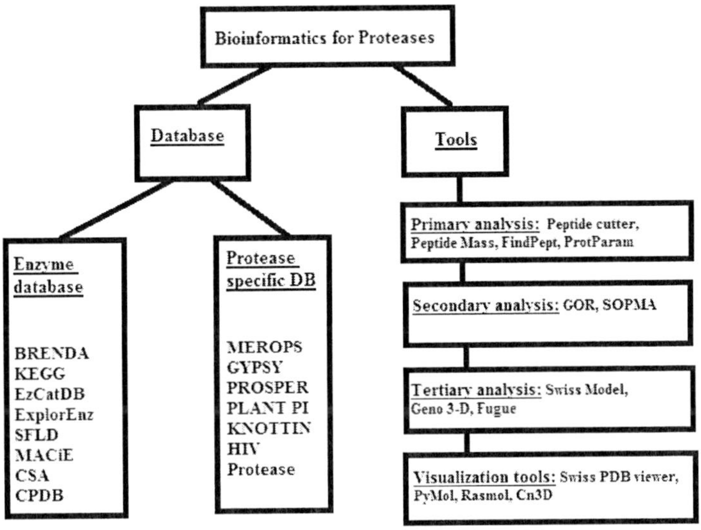

Figure 3. Bioinformatics resources for proteases.

2. DATABASES

Various bioinformatics databases can be used for protease analysis. These may or may not be protease specific. Different types of enzyme databases can also be used to study their metabolic pathways and mechanism of action. Thus, databases for proteases may fall into the category of enzyme databases and protease specific databases.

2.1. Enzyme Databases

Enzymes have a variety of biotechnological applications due to their activity of catalyzing chemical reactions and their involvement in metabolism of various organisms.

Enzymatic databases provide information regarding the source, occurrence, properties, functions, kinetics and metabolism of enzymes. Proteases are also enzymes and hence, a number of enzymatic databases can be used for protease analysis.

2.1.1. BRENDA

BRENDA (https://www.brenda-enzymes.org/) is a database that stores information regarding enzymes that have been obtained manually from literature studies and other data that has been obtained via automated methods such as text mining. This is a web based user interface that enables an efficient access to the enzymatic data. BRENDA was originally developed in the year 1987 at the German National Centre for Biotechnology now called as the Helmholtz Centre for Infection Research in Braunschweig and was actually published as a series of book. Its name is an abbreviated form for Braunschweig Enzyme Database.

It consists of 40 data fields and more. These data fields consist of enzymatic information on about 7,000EC numbers that are categorized based on IUBMB. The enzymatic information includes enzyme's nomenclature, specificity and reaction, structure of the enzyme, its isolation, preparation and stability. It also takes into account the Km value and turnover number along with other kinetic parameters, mutations and genetically engineered enzymes, their applications and ligand binding data. At present, the information stored in BRENDA has been derived from 140,000 different scientific articles. All the enzyme entries in the database are linked to at least one reference from the literature, the source and origin of the enzyme, the protein sequence and its availability. One important feature of BRENDA is that it represents 110000 ligands of enzymes that are present as names, synonyms or as chemical structure. The low molecular weight compounds that are capable of interacting with enzymes are termed as "ligands". The ligands may be co-substrate or cofactors, inhibitors, metal ions or primary metabolites. Such molecules are usually derived from antibiotics that occur naturally in nature or other man-made compounds that have been chemically synthesized for the development of drugs or other pesticides [11].

2.1.2. KEGG

KEGG is expanded as Kyoto Encyclopaedia of Genes and Genomes (https://www.genome.jp/kegg/). It is a collection of databases that has information related to genomes, biological processes and pathways, diseases, drugs and other chemical compounds. This database has been used for a number of bioinformatics applications that include analysis of genomic data, metagenomics, metabolomics and other omics studies.

KEGG utilizes computers to represent the biological system. It makes use of building blocks and wiring systems. It combines the genes, small molecules and chemical reactions that act as building blocks along with molecular interactions and reaction networks that act as wiring systems. This concept is the basis of KEGG [12].

2.1.3. ExplorEnz

ExplorEnz (http://www.enzyme-database.org/) is a database that acts as a primary container for EC numbers and enzyme-specific data which is collected and gathered on behalf of the IUBMB. This database was developed at Trinity College Dublin in the year 2005. The role of this database is to provide access to the data of IUBMB Enzyme Nomenclature List that has been incorporated in various other bioinformatics databases such as BRENDA, KEGG, etc. The formatting of chemical names is as per the IUPAC guidelines and the data is stored in MySQL. Andrew McDonald is the person behind ExploEnz [13].

2.1.4. EzCatDB

The EzCatDB database (http://ezcatdb.cbrc.jp/EzCatDB/) contains information on enzymes that have been manually classified depending upon their active site structures and their mechanism to catalyze reactions. This information is obtained from literature, amino acid sequences from UniProtKB and their 3D structures from PDB (Protein Data Bank). Various types of reactions such as transfer, hydrolysis, elimination, addition, isomerization, electron transfer, etc. can be classified. It provides information about ligand molecules that are present within the structures of enzymes in the PDB. These can be present in the form of co factors, substrates, products and intermediates [14].

2.1.5. Structure Function Linkage Database

The Structure-Function Linkage Database (SFLD, http://sfld.rbvi.ucsf.edu/) is used to determine structure-function relationship for a large number of functional enzyme superfamilies. These superfamilies come from a common predecessor and their active site domains are conserved where as they catalyze wide variety of reactions. Therefore, the user may get confused since these enzymes look alike in spite of their difference in functions. To overcome this complexity, members of the super family are subdivided into various subgroups based on information from sequences and families that consist of enzymes which use the same mechanistic method to catalyze the reaction. Thus, SFLD is an efficient database that enables the user to access all these levels by providing a number of browsing and searching options. The superfamily sets present in the database are classified either manually or automatically using various browsing and download options for all hierarchical levels. Other important features of the database include MSA (Multiple Sequence Alignment), sequence similarity networks, etc. This helps in visualization of functional trends mapped as per the sequence similarity [15].

2.1.6. MACiE

MACiE is expanded as Mechanism, Annotation and Classification in Enzyme (https://bio.tools/macie). This database is freely available and the primary focus of this database is to help us study the evolution of the catalytic mechanism of the enzymes. It not only provides information on the overall reaction but it also represents the catalytic steps of an enzyme reaction [16].

2.1.7. Catalytic Site Atlas

CSA stands for Catalytic Site Atlas (http://www.ebi.ac.uk/thornton-srv/databases/CSA) provides information about the enzymes stored in PDB. It consists of annotations for catalytic residues of enzymes and the annotated sites. These include an original hand annotated set and an extra homologous set. The former contains information that is derived from literature source that assigns catalytic residues based on a definite criterion whereas the latter contains information derived from PSI-BLAST and sequence alignments. The queries in CSA are given using EC numbers, PDB codes or Swiss-Prot identifiers. 30% of the EC numbers present in PDB are covered by CSA Version 1.0 db. This version contains 177 originals manually annotated entries and 2608 homologous entries. Hence, this database is updated regularly [17].

2.1.8. CPDB

CPDB (http://consensuspathdb.org/) stands for Consensus Pathway Database. This database integrates various types of interaction networks found in human beings. These include protein-protein interaction networks, genetic, metabolic, signaling and other biochemical pathways. It was developed in the year 2008 and it is freely accessible [18].

2.2. Protease Specific Databases

As the name suggests, this database is specially designed to deal with the protease analysis.

2.2.1. MEROPS

The MEROPS database (https://www.ebi.ac.uk/merops/) provides information about proteases, their inhibitors and the substrates. It was developed in the year 1996. This database is based on a hierarchical classification system of proteases in which homologous sets of proteases and its inhibitors are assembled as protein species. These protein species are further categorized into families which can be further grouped into clans. This kind of classification helps to relate information at each level. An important characteristic of the database is that it helps distinguish proteases based on their ability to cleave a particular substrate and their interaction with inhibitors. Number of complete genomes is greater than the number of new protease families and its clans. A special analytical tool has been added that indicates the increase or decrease of protease genes relative to related species. The database categorizes information in protease family, clan and identifier [19].

2.2.2. Gypsy Database

The GyDB (http://gydb.org/index.php/Main_Page) project aims to categorize the non-redundant of various genes based on their evolutionary relationships. The GyDB project has been established into a database. The very first release of the database was based upon establishing an evolutionary relationship between Ty3/Gypsy and Retroviridae LTR retro elements and clan amino acids of Aspartic Proteases [20].

2.2.3. PROSPER

PROSPER can be expanded as Protease Specificity Prediction Server (https://prosper.erc.monash.edu.au/home.html). This database is used to identify unique substrates and active sites of 24 different peptidase families. It uses protein sequences and structural information for this purpose. It involves two approaches in this process, one is the support vector regression approach and other is the bi-profile Bayesian feature extraction approach. The software is capable of determining active sites of various proteases using different machine learning techniques to which a particular substrate can bind. Thus, it is very useful in *in-silico* identification tool for identifying active sites of proteases. PROSPER database works in three broad steps: the input step, the prediction step and the output step [21].

2.2.4. PLANT PI

PLANT PI (http://plantpis.ba.itb.cnr.it/) stands for Plant Protease Inhibitors. As the name suggests, this database is specially designed to deal with information related to plant inhibitors and genes related to it. For every protease inhibitor in the database, reference links containing sequence information and functional properties are provided that link to other databases. Currently the database stores information on 351 plant protease inhibitors. The categorization of PI families is the bases of organizing annotations of PIs in a database. The PI families are reported as per the InterPro which is Integrated Resource of Protein Domains and Functional Sites. The classification depends upon the primary amino acid sequence [22].

2.2.5. KNOTTIN Database

Knottins or the inhibitor cystine knots are those mini proteins that are distinguishable due to the presence of 3 intertwined disulphide bridges that leads to the formation of a pseudo knot. These are ultra-stable in nature and are widely used for drug designing and medical imaging purposes.

Thus, KNOTTIN database (www.dsimb.inserm.fr/KNOTTIN/) was designed to provide information on knottin sequences, structure, functions and evolution. It also contains a bibliographic data and other computational tools specific for knottins [23].

2.2.6. HIV Protease Database

The HIV Protease Database contains structural information on HIV proteases. It consists of experimentally derived three dimensional structures of HIV type1 (Human immunodeficiency virus type 1), HIV type2 (Human immunodeficiency virus type2), SIV (Simian immunodeficiency virus) proteases and their complexes with inhibitors and other specific substrates.

This database is a great tool for researchers who aim to develop a drug or a therapy against AIDS (Acquired Immunodeficiency Syndrome) [24].

2.3. Tools for Protease Analysis

The tools for protease analysis are categorized in four *i.e.*, primary, secondary and tertiary analysis and visualization.

2.3.1. Primary Analysis

The tools used for primary analysis of protease are as follows:

2.3.1.1. Peptide Cutter

Peptide Cutter (https://web.expasy.org/peptide_cutter/) is a tool that is used for identification of cleavage sites in a protease molecule. It is used to search an amino acid sequence from SWISS-PROT and/or TrEMBL databases. It may also search for user entered protein sequence. The tool can perform analysis for a single protease as well as for multiple proteases [25].

2.3.1.2. Peptide Mass

Peptide Mass (https://web.expasy.org/peptide_mass/) generates masses of a cleaved peptide with the help of a protease that is provided by the UniProt database or is entered by the user. It also provides information about the isoelectric point and peptide mass for the protein of interest. The information generated on masses can be used by the user for any PTM (Post Translational Modifications) or it may focus on proteins whose masses may be affected by polymorphism or variants [26].

2.3.1.3. Multident

It is a tool (https://web.expasy.org/multiident/) used to determine proteins from 2-Dimensional Gels. A number of properties lead to identification of proteins such as protein PI (isoelectric point), molecular mass of proteins estimated using mass spectrometry/2-D Gels, source from which the protein is originated and composition of amino acid sequence.

This database is designed to accommodate various combinations in order to achieve definite identity of protein identification. The program eventually works in a modular fashion. The current version of this database first generates a list of similar proteins along with their amino acid composition. Further, it utilizes other protein parameters such as isoelectric point, molecular weight, sequence tag, peptide mass data in order to identify the desired protein [27].

2.3.1.4. iPEP

Peptide based proteomics focus on quantification and identification of post translational modifications in proteins that are isolated from biological samples. It uses bottom up approach to cleave protein sequences via high throughput technologies such as liquid chromatography and mass spectrometry.

iPEP (http://ipep.moffitt.org) is a web based tool specifically designed for comparison between different protease digests in order to determine their effectiveness in identifying specific sequences. Optimization can also take place using this tool for analyzing certain group of proteins related to the proteome. It provides information regarding the protease sequence, their molecular weights and functions using Gene Ontology. It also utilizes a number of filters to limit the output size and compare the digested peptides. Prosite format is used to determine the consensus sites and residues in Post Translational Modifications [28].

2.3.1.5. FindPept

FindPept (http://www.expasy.org/tools/findpept.html) is a tool that uses Peptide Mass fingerprinting for identification of peptide masses. It is used to determine masses from inadequate cleavage such as missed cleavage sites, unspecified proteolytic cleavage or from contaminated keratins. It also considers post translational modifications that is obtained from SWISS-PROT database or given by the user and also considers chemical modifications in a protein [29].

2.3.1.6. ProtParam

TheExpsyProtParam (https://web.expasy.org/protparam/) is a tool that has been developed to determine the physicochemical parameters of proteins. Proteases are protein molecules and hence various physico-chemical properties of proteases can be computed using this tool. The user can input the protease sequence from UniProt/SwissProt/TrEMBL or any other source in FASTA format or provide the unique accession number or ID. A lot of information can be derived from the primary sequence such as total number of amino acids, isoelectric point, GRAVY (Grand Average Hydropathy), molecular weight, etc. [25].

2.3.1.7. PSORT

PSORT refers to Protein Subcellular Localization Prediction Tool (https://www.genscript.com/psort.html). This tool is used for the identification of subcellular localization sites in a protein sequence of a Gram Negative Bacteria. This tool predicts the location of a protein inside a cell.

This database converts the primary sequence of protein into numerical localization features depending upon sorting signals, composition of the sequence and its functional motifs [30].

2.3.2. Secondary Analysis

The analysis of secondary elements of protease sequence can be done by tools as follows:

2.3.2.1. GOR

GOR stands for Garnier-Osguthorpe-Robson (http://cib.cf.ocha.ac.jp/bitool/GOR/). It is named after its developers. This tool is used for the prediction of secondary structure of proteins. It was developed after the Chou-Fasman method. It makes use of a propensity scale to calculate the alpha helix, beta sheets, turns and random coils. It also considers the interaction of side chain residues of neighboring amino acids during calculation of propensity value [31].

2.3.2.2. SOPMA

SOPMA means Self Optimized Prediction Method with Alignment (https://npsaprabi.ibcp.fr/cgibin/npsa_automat.pl?page=/NPSA/npsa_sopma.html). It is a secondary structure prediction method. It is based on the nearest neighbour method and was developed in the year 1995[32].

2.3.3. Tertiary Analysis

The tools used for tertiary structure analysis of protease based on various approaches are as follows:

2.3.3.1. Swiss Model

Swiss Model makes use of Homology modelling to predict tertiary structure of protein (https://swissmodel.expasy.org/). It is the most appropriate method for the generation of three dimensional structures of proteins. It uses templates as reference to determine the co-ordinates of query sequences [33].

2.3.3.2. Geno 3-D

Just like Swiss Model, Geno 3-D (https://geno3d-prabi.ibcp.fr/cgi-bin/geno3d_automat.pl?page=/GENO3D/geno3d_home.html) is an automatic web server used for tertiary structure prediction of proteins. This program also makes use of homology modelling for 3-D structure prediction.

2.3.3.3. Fugue

Fugue (http://mizuguchilab.org/fugue/) is a program used for the recognition of distantly related homologous sequences by sequence-structure comparison. Depending upon the query sequence, fugue scans a database for structure information and calculates the sequence-structure compatibility score. The program then displays a list of potential homologues and their alignments.

2.3.4. Visualisation Tool

These are the tools that are used to visualize the structural information regarding protease (Table 1).

Table 1. Various visualization tools for structural analysis

S.No.	Visualization tool	Link
1	SwissPDB-viewer	https://spdbv.vital-it.ch/
2	Pymol	https://pymol.org/2/
3	Cn3D	https://www.ncbi.nlm.nih.gov/Structure/CN3D/cn3dwin.shtml
4	Rasmol	http://www.openrasmol.org/

2.3.4.1. SwissPDB-Viewer

SwissPDB-viewer is a visualization tool used for the analysis of several proteins at the same time. It is used to determine structural alignments by superimposing the two proteins. It is also used for comparing the active sites and other relevant parts. This tool was developed in the year 1994 and is popularly used for visualization purposes [34].

2.3.4.2. PyMOL

PyMOL is a computer software which is created by Warren Lyford DeLano. It is a visualization system that is most commonly used in structural biology. This software is used to create high quality 3D images of various micro and macro biological molecules especially proteins. This software is commercialized by Schrodinger and uses Python for its functioning [35].

2.3.4.3. Cn3D

This tool allows us to view 3-dimensional structure from NCBI's Entrez Structure Database. It is available for Windows, Macintosh and Unix. It can be used to display sequence, structure and alignment [36].

2.3.4.4. Rasmol

RasMol is used for visualization of structures of biological macromolecules. This was developed in the year 1990 by Roger Sayle [37].

REFERENCES

[1] Dayanand, C.D. (2013). Evaluation of comparative total proteolytic activity in plant lattices. *International Journal of Life Sciences Biotechnology and Pharma Research*, 2:47-55.

[2] Gupta, R., Beg, Q. and Lorenz, P. (2002). Bacterial alkaline proteases: Molecular approaches and industrial applications. *Applied Microbiology and Biotechnology,* 59(1):15-32.

[3] Chang, H.Y. and Yang, X. (2000). Proteases for cell suicide: Functions and regulation of caspases. *Microbiology and Molecular Biology Reviews*, 64(4):821-846.

[4] Saibi, W., Zouari, N., Masmoudi, K. and Brini, F. (2016). Role of the durum wheat dehydrin in the function of proteases conferring salinity tolerance in *Arabidopsis thaliana* transgenic lines. *International Journal of Biological Macromolecules,* 85:311-316.

[5] Smid, I., Rotter, A., Gruden, K., Brzin, J., Gasparic, M.B., Kos, J., Zel, J. and Sabotic, J. (2015). Clitocypin, a fungal cysteine protease inhibitor, exerts its insecticidal effect on Colorado potato beetle larvae by inhibiting their digestive cysteine proteases. *Pesticide Biochemistry and Physiology*, 122:59-66.

[6] Singh, R., Kumar, M., Mittal, A. and Mehta, P.K. (2016). Microbial enzymes: Industrial progress in 21st century. *3 Biotech.* 6(2):1-15.

[7] Verbovsek, U., Noorden, C.J.V and Lah, T.T. (2015). Complexity of cancer protease biology: Cathepsin K expression and function in cancer progression. *Seminars in Cancer Biology*, 35:71-84.

[8] Rani, K., Rana, R. and Datt, S. (2012). Review on latest overview of proteases. *International Journal of Current Life Science,* 2(1):12-18.

[9] Lesk, A.M. (2002). *Introduction to Bioinformatics.* Oxford University Press, New York.

[10] Alderson, R.G., Ferrari, L.D., Mavridis, L., McDonagh, J.L., Mitchell, J.B.O. and Nath, N. (2012). Enzyme Informatics. *Current Topics in Medicinal Chemistry*, 12(17):1911-1923.

[11] Schomburg, I., Chang, A. and Schomburg, D. (2002). Enzyme data and metabolic information. *Nucleic Acids Research*, 30(1):47-49.

[12] Kanehisa, M. and Goto, S. (2000). KEGG: Kyoto Encyclopedia of Genes and Genomes. *Nucleic Acids Research*, 28(1): 27-30.

[13] McDonald, A.G., Boyce, S. and Tipton, K.F. (2009). ExplorEnz: the primary source of the IUBMB enzyme list. *Nucleic Acids Research*, 37(1):593-597.

[14] Nagano, N. (2005). EzCatDB: the Enzyme Catalytic-mechanism Database. *Nucleic Acids Research*, 33:407-412.

[15] Akiva, E., Brown, S., Almonacid, D.E., Barber, A.E., Custer, A.F., Hicks, M.A., Huang, C.C., Lauck, F., Mashiyama, S.T., Meng, E.C., Mischel, D., Morris, J.H., Ojha, S., Schnoes, A.M., Stryke, D., Yunes, J.M., Ferrin, T.E., Holliday, G.L. and Babbitt, P.C. (2014). The Structure-Function Linkage Database. *Nucleic Acids Research*, 42:521-530.

[16] Holliday, G.L., Bartlett, G.J., Almonacid, D.E., Boyle, N.M.O, Rust, P.M., Thornton, J.M. and Mitchell, J.B.O. (2005). MACiE: A Database of Enzyme Reaction Mechanisms. *Bioinformatics*, 21(23):4315-4316.

[17] Porter, C.T., Bartlett, G.J. and Thornton, J.M. (2004). The Catalytic Site Atlas: a resource of catalytic sites and residues identified in enzymes using structural data. *Nucleic Acids Research*, 32:129-133.

[18] Kamburov, A., Stelzl, U., Lehrach, H. and Herwig, R. (2012). The Consensus PathDB interaction database: 2013 update. *Nucleic Acids Research*, 41(D1):793-800.

[19] Rawlings, N.D., Barrett, A.J. and Bateman, A. (2010). MEROPS: the peptidase database" *Nucleic Acids Research*, 38:227-233.

[20] Lorens, C., Futami, R., Covelli, L., Dominguez, L.E., Viu, J.M., Tamarit, D., Rodriguez, J.A., Ripolles, M.V., Fuster, G., Bernet, G.P., Maumus, F., Pomer, A.M., Sempere, J.M., LaTorre A. and Moya, A. (2011). The Gypsy Database (GyDB) of Mobile Genetic Elements. *Nucleic Acids Research,* 39:70-74.

[21] Song, J., Tan, H., Perry, A.J., Akutsu, T., Webb, G.I., Whisstock, J.C. and Pike, R.N. (2012). PROSPER: an integrated feature-based tool for predicting protease substrate cleavage sites. *PloS One*, 7(11):e50300.

[22] Leo, F.D., Volpicella, M., Licciulli, F., Liuni, S., Gallerani, R. and Ceci, L.R. (2002). PLANT-PIs: a database for plant protease inhibitors and their genes. *Nucleic Acids Research*, 30(1):347-348.

[23] Postic, G., Gracy, J., Périn, C., Chiche, L. and Gelly, J.C. (2018). KNOTTIN: The database of inhibitor cystine knot scaffold after 10 years, toward a systematic structure modeling. *Nucleic Acids Research*, 46:454–458.

[24] Ravichandran, V., Vondrasek, J., Gilliland, G., Bhat, T.N. and Wlodawer, A. (2002). HIV protease structural database Proc. *IEEE Computer Society Bioinformatics Conference.*

[25] Gasteiger, E., Hoogland, C., Gattiker, A., Wilkins, M.R. and Appel, R.D., Bairoch, A. (2005). Protein identification and analysis tools on the ExPASy server. In Walker J.M. (eds), *The Proteomics Protocols Handbook*. Humana Press, 571-607.

[26] Wilkins, M. R., Lindskog, I., Gasteiger, E., Bairoch, A., Sanchez, J. C., Hochstrasser, D. F., and Appel, R. D. (1997). Detailed peptide characterization using PEPTIDEMASS–a World-Wide-Web-accessible tool. *Electrophoresis*, *18*(3-4): 403-408.

[27] Wilkins, M.R., Gasteiger, E., Wheeler, C.H., Lindskog, I., Sanchez, J.C., Bairoch, A., Appel, R.D., Dunn, M.J. and Hochstrasser, D.F. (1998). Multiple parameter cross-species protein identification using MultiIdent-a world-wide web accessible tool. *Electrophoresis*, *19*(18):3199-3206.

[28] Lu, D., Liu, R.Z., Izumi, V., Fenstermacher, D., Haura, E.B., Koomen, J. and Eschrich, S.A. (2008). IPEP: an *in silico* tool to examine proteolytic peptides for mass spectrometry. *Bioinformatics*, 24(23):2801-2802.

[29] Gattiker, A., Bienvenut, W.V., Bairoch, A. and Gasteiger, E. (2002). FindPept, a tool to identify unmatched masses in peptide mass fingerprinting protein identification. *Proteomics*, 2:1435-1444.

[30] Nakai, K., and Horton, P. (1999). PSORT: a program for detecting sorting signals in proteins and predicting their subcellular localization. *Trends in Biochemical Sciences*, *24*(1): 34-35.

[31] Garnier, J., Gibrat, J.F. and Robson, B. (1996). GOR method for predicting protein secondary structure from amino acid sequence. *Methods in Enzymology*, 266:540-553.

[32] Geourjon, C., and Deleage, G. (1995). SOPMA: significant improvements in protein secondary structure prediction by consensus prediction from multiple alignments. *Bioinformatics*, *11*(6):681-684.

[33] Schwede, T., Kopp, J., Guex, N. and Peitsch, MC. (2003). SWISS-MODEL: an automated protein homology-modeling server. *Nucleic Acids Research,* 31:3381-3385.

[34] Waterhouse, A., Bertoni, M., Bienert, S., Studer, G., Tauriello, G., Gumienny, R., Heer, F.T., de Beer, T.A.P., Rempfer, C., Bordoli, L. and Lepore, R. (2018). SWISS-MODEL: homology modelling of protein structures and complexes. *Nucleic acids research, 46*(W1):296-303.

[35] DeLano, W.L. (2002). *PyMOL.* DeLano Scientific LLC, San Carlos, CA, 700.

[36] Wang, Y., Geer, L. Y., Chappey, C., Kans, J. A., and Bryant, S. H. (2000). Cn3D: sequence and structure views for Entrez. *Trends in biochemical sciences*, *25*(6):300-302.

[37] Sayle, R. A., and Milner-White, E. J. (1995). RASMOL: biomolecular graphics for all. *Trends in biochemical sciences*, *20*(9):374-376.

ABOUT THE EDITOR

Nelson Pérez Guerra
Professor/Researcher, Department of Analytical and Food Chemistry.
Food Science and Technology Faculty, University of Vigo,
Ourense Campus, Ourense, Spain
Fax +34-988-387001
E-mail: nelsonpg@uvigo.es

Dr. Nelson Pérez Guerra was born on September 30, 1996 in Santiago de Cuba, Cuba. He was graduated as a Chemical Engineer from the ISPJAE-University of Havana in Cuba in 1989. He started working as a Production Engineer at the Flour and Canned Vegetables Companies (both in Cuba) from 1989 to 1991. From 1991 to 1996, he worked as an Assistant Professor of Biochemistry, Microbiology and Process Engineering I and II at the University of Oriente (Santiago de Cuba). He graduated as a Doctor of Food and Technology Science on 1st October 1999. From 1999 to date, he works as a Full Professor of Biochemistry at Vigo University.

He has published 17 chapters in different International Books (was the scientific editor in one International Book), more than 50 papers in peer-reviewed scientific journals and 39 proceedings in National and International Congresses. He is a member of the Editorial Board of 14 international scientific journals and serves as scientific reviewer for different international journals.

He has supervised 4 doctoral theses, 4 graduate theses, 40 end-of-course projects and 9 master's final projects. He has participated in more than 30 research projects (principal investigator in 7 of them).

INDEX

A

D

E

F

G

H

I

J

K

L

M

N

O

P

R

S

T

V

W

Y